Maren Reimold

The Newtonian Limit of General Relativity

Maren Reimold

The Newtonian Limit of General Relativity

Südwestdeutscher Verlag für Hochschulschriften

Imprint

Any brand names and product names mentioned in this book are subject to trademark, brand or patent protection and are trademarks or registered trademarks of their respective holders. The use of brand names, product names, common names, trade names, product descriptions etc. even without a particular marking in this work is in no way to be construed to mean that such names may be regarded as unrestricted in respect of trademark and brand protection legislation and could thus be used by anyone.

Publisher:
Südwestdeutscher Verlag für Hochschulschriften
is a trademark of
Dodo Books Indian Ocean Ltd., member of the OmniScriptum S.R.L Publishing group
str. A.Russo 15, of. 61, Chisinau-2068, Republic of Moldova Europe
Printed at: see last page
ISBN: 978-3-8381-2528-2

Zugl. / Approved by: Tübingen, Eberhard Karls Universität, Diss., 2010

Copyright © Maren Reimold
Copyright © 2011 Dodo Books Indian Ocean Ltd., member of the OmniScriptum S.R.L Publishing group

Contents

Introduction   **3**
    0.1   Transitions from tangent and cotangent space and induced connections   6
    0.2   Concepts of curvature   7
    0.3   Newton's theory of gravitation and Einstein's theory of relativity   12

1   Frame theory   **17**
    1.1   The structure of the frame theory   17
    1.2   Linear Algebra   21
    1.3   Transfer to the frame theory   28
    1.4   The case $\lambda = 0$ 32

2   The Newtonian limit: definition and existence   **61**
    2.1   Definition of the Newtonian limit   61
    2.2   Extension of spacetimes   64
    2.3   Examples   64
    2.4   Existence of a limit   74
    2.5   Static and spherically symmetric spacetimes   80

3   Existence of genuine Newtonian limits   **93**
    3.1   Transformation of coordinates   94
    3.2   Conditions for the curvature tensor   102
    3.3   Asymptotically flat spacetimes   105

A   Appendix   **121**
    A.1   The Schwarzschild spacetime   121
    A.2   The Kerr spacetime   127

Index   **147**

Bibliography   **149**

Introduction

Since the theory of General Relativity has existed, the corresponding question has been discussed as to whether and to what degree Newton's theory of gravitation can be considered as a special case or, at least, as a limit situation of General Relativity. Already on 18th November 1915, one week after presenting his field equations at the "Preußische Akademie der Wissenschaften" in Berlin, Albert Einstein wrote in a letter to David Hilbert that he had no problem to find a covariant formulation for his field equations. With the Riemannian curvature tensor he had a tensor field at his disposal which provided a sufficient degree of generality. In fact, however, he had problems to formulate a field equation which contained the Newtonian gravitation as a special case:

> *Die Schwierigkeit bestand nicht darin allgemein kovariante Gleichungen für die $g_{\mu\nu}$ zu finden; denn dies gelingt leicht mit Hilfe des Riemann'schen Tensors. Sondern schwer war es, zu erkennen, dass diese Gleichung eine Verallgemeinerung des Newton'schen Gesetzes bilden. Dies gelang mir erst in den letzten Wochen (...), während ich die einzig möglichen allgemein kovarianten Gleichungen, [die] sich jetzt als die richtigen erweisen, schon vor 3 Jahren mit meinem Freund Grossmann in Erwägung gezogen hatte. Nur schweren Herzens trennten wir uns davon, weil mir die physikalische Diskussion scheinbar ihre Unvereinbarkeit mit Newtons Gesetz ergeben hätte.*
>
> [39], Document 148, p. 201

Thus, already Albert Einstein was interested in the question to which extend the Newtonian theory can be regarded as a limit situation of the theory of Relativity. In his work he studied this problem, but his considerations were limited to the special case of the Schwarzschild spacetime and he used very special coordinates. He succeeded in detecting the Newtonian gravitation field in this special coordinate system by considering $c \to \infty$, where c is the speed of light. The transition of the Einstein gravitation potential to the Newtonian one remained a little bit obscure.

In the 1920s, E. Cartan took a very important step to a better understanding of this limit process. He formulated the Newtonian theory as a field theory on a four-dimensional manifold with a connection ∇ as gravitation field, see [4] and [5].

In the 1980s, Jürgen Ehlers developed his frame theory and gave an answer to the question to which extend the transition from Einstein's theory of relativity to Newton's theory of gravitation was understood. Research into the implications of this frame theory were continued by Martin Lottermoser among others.

But why is it interesting to understand the connection between these two theories? For instance, one of the reasons is that, in general, we are interested in unifying successful theories having the same purpose as in this case. Furthermore, it seems to be helpful to have the possibility of transferring results from one theory to the other. This could be of some importance to people who are interested in a better understanding of post-Newtonian approximations, for example. During my work on the Newtonian limit of General Relativity I met a lot of people, especially physicists who did not understand why I was interested in this purpose, since they considered the limit relation between the two theories as already known and well understood. Such views already appear in the papers of Jürgen Ehlers:

> Obwohl manche Lehrbücher und Monographien den Eindruck erwecken, als sei die genannte Grenzwertbeziehung ($\to$) eine wohlverstandene Sache, ist mir aus der Literatur kein exakt formulierter und bewiesener Satz bekannt, der diese Meinung rechtfertigen würde. Es ist zwar leicht möglich und zur vorläufigen Orientierung auch nützlich, aus den Grundgleichungen der Einsteinschen Theorie formal diejenigen der Newtonschen Theorie zu gewinnen, indem man einen geeigneten Ansatz für die Metrik und die Materievariablen macht und kleine Größen vernachlässigt, d.h. durch Null ersetzt ([...]). Eine mathematische Rechtfertigung solcher heuristischer "Ableitungen" steht aber noch aus.

[10], pp. 65-66

With his frame theory, Jürgen Ehlers took the decisive step towards understanding this limit connection. The mathematical objects of his frame theory are the so called *Ehlers spacetimes* $M_\lambda = (M, g(\lambda), h(\lambda), \nabla(\lambda), T(\lambda), \lambda)$, where M is a four-dimensional manifold, $g(\lambda)$ and $h(\lambda)$ are metrics, $\nabla(\lambda)$ a connection on M, $T(\lambda)$ a matter tensor and $\lambda \in \mathbb{R}$ a real number. These objects have to satisfy some axioms. Then it can be shown that for $\lambda > 0$, $-g$ is a Lorentzian metric, ∇ its Levi-Civita connection and Einstein's field equations hold. Therefore, for $\lambda > 0$ and λ fixed, an Ehlers spacetime is an *Einstein spacetime*, which means a spacetime in the sense of General Relativity.

While this can easily be seen, it is not obvious that the case of $\lambda = 0$ contains models of Newton's theory of gravitation. For instance, at the outset the frame theory starts with a four-dimensional manifold and we have to discuss why this manifold results in a Newtonian space for $\lambda = 0$. But we will see that this is the case under certain circumstances.

Up to this moment, Ehlers frame theory has been well-known, (see, for instance, [31]), but not so in a satisfactory way for global differential geometers. Therefore, one aim of this work is to explain the global structure in case of $\lambda = 0$, as there are some unsolved problems in this context.

We start this work by introducing some notations and discussing basic concepts. In the first chapter we then examine the frame theory of Jürgen Ehlers. After introducing the mathematical objects and the axioms of the frame theory, we explain how the theory of General Relativity and Newton's theory of gravitation can be detected within the frame theory. We will have to do into detail discussing the case of $\lambda = 0$, where Newton's theory may be involved. In this context we can present some new results, for example, we are able to show that the manifold has a global product structure under certain circumstances (see for instance (1.4.10) and (1.4.32)) or that there is a global timelike vector field on M (see (1.4.16)).

In the second chapter we will talk about the concept of a Newtonian limit. We start with its definition and discuss how we can extend a solution of Einstein's field equations to a family of Ehlers spacetimes. In other words, we adopt an approach opposed to the one in the first chapter: We start with an Einstein spacetime and demonstrate how it can be included into the frame theory. Furthermore, we will talk about some standard examples of Newtonian limits and discuss them in detail. Then we reconsider the question under which condition a family of Ehlers spacetimes has a Newtonian or at least a quasi-Newtonian limit. These conditions are partially known (see, for instance, [31] and [10]), but we are able to give a more geometrical condition for the existence of the limit of the metrics (see (2.4.5)). Finally, we will show how to find a Newtonian limit for spherically symmetric and static solutions of Einstein's field equations (see (2.5)).

In the previous paragraph the concept of a quasi-Newtonian limit has been mentioned. In the course of this work we will explain that for $\lambda = 0$ sometimes not genuine Newtonian cases appear. In particular, there is a time-dependent vector field v on $\mathbb{R}^3$ which is an obstacle to a limit to be a genuine Newtonian one. In the third chapter, we therefore discuss conditions which ensure that this vector field v disappears. First, we discuss its transformation behaviour and show that it can be transformed to zero if it only depends on time. Then we discuss some conditions of Ehlers and Trautman for the curvature tensor. Already Ehlers and Lottermoser mentioned that a right definition of isolated systems, which means of asymptotically flat spacetimes, might provide a condition for the existence of a genuine Newtonian limit. Here we introduce a definition of asymptotically flat families of Ehlers spacetimes and show that, if these families have a quasi-Newtonian limit, they have a genuine Newtonian one (see (3.3.8) and (3.3.10)). Furthermore, we show that our standard examples satisfy the definition of asymptotically flat spacetimes, which demonstrates that our definition is reasonable.

In the appendix, we list calculations about the extended Schwarzschild and Kerr solutions.

I want to give special thanks to Prof. Dr. Frank Loose for help and motivation, especially in times when it was hard to carry on. Furthermore, thanks to my very dear colleagues at the Eberhard Karls Univeristy of Tübingen. Of course, I have to thank my family for being always there for me, particularly thanks to my parents and my husband for motivating and supporting me all the time.

In this part we will introduce basic concepts and notations used in this work. At the end of this section we will discuss some aspects of Newton's theory of gravitation and Einstein's theory of relativity.

In the following paragraphs let M always be a smooth manifold.

0.1 Transitions from tangent and cotangent space and induced connections

Notation 0.1.1 Let $\pi : E \to M$ be a smooth vector bundle. We denote the *set of all global sections* by
$$\Gamma(M; E) := \{s : M \to E \text{ smooth} : \pi \circ s = \mathrm{id}_M\}.$$
If $E = TM$ or $E = TM^*$, we denote the sections (which are vector fields or differential forms respectively) by
$$\mathfrak{X}(M) = \Gamma(M; TM)$$
and
$$\mathcal{E}^{(1)}(M) = \Gamma(M; TM^*).$$

Notation 0.1.2 Let $g \in \Gamma(M; T^{(0,2)}M)$ and $h \in \Gamma(M; T^{(2,0)}M)$. Then we define
$$\varphi_p : TM_p \to TM_p^*, \ \xi \mapsto (\eta \mapsto g_p(\xi, \eta))$$
$$\psi_p : TM_p^* \to TM_p, \ \alpha \mapsto (\beta \mapsto h_p(\alpha, \beta)) \in TM_p^{**} = TM_p$$
as the *transitions from the tangent space to the cotangent space and vice versa induced by g and h*.

In order to have a better understanding of the maps φ_p and ψ_p, note:
$$\langle \varphi_p(\xi), \eta \rangle = g_p(\xi, \eta), \ \text{for } \xi, \eta \in TM_p,$$
and
$$\langle \beta, \psi_p(\alpha) \rangle = h_p(\alpha, \beta), \ \text{for } \alpha, \beta \in TM_p^*,$$
where
$$\langle \_, \_ \rangle : TM_p^* \times TM_p \to \mathbb{R}, \ \langle \alpha, \xi \rangle = \alpha(\xi)$$
denotes the *natural pairing between covectors and tangent vectors*.

With the help of these transitions we are able to "transform" tensor fields: if we use the index "b" we apply φ, if we use ψ we denote "#". For example, for $T \in \Gamma(T^{(2,0)}M)$, we have $T^{bb} \in \Gamma(T^{(0,2)}M)$.

Remark 0.1.3 If you have a look at (0.1.2) you see that we write
$$TM_p^{**} = TM_p.$$

We want to point out that we can identify TM_p^{**} with TM_p with the help of the following canonical isomorphism:
$$\iota_p : TM_p \to TM_p^{**}, \ \iota_p(\xi)(\alpha) = \langle \alpha, \xi \rangle,$$
for $\xi \in TM_p$ and $\alpha \in TM_p^*$, where
$$\langle \_,\_ \rangle : TM_p^* \times TM_p \to \mathbb{R}, \ \langle \alpha, \xi \rangle = \alpha(\xi)$$
denotes the natural pairing between covectors and tangent vectors. This means that for all $\alpha \in TM_p^*$
$$\psi_p(\alpha) = \xi \Leftrightarrow h_p(\alpha, \beta) = \langle \beta, \xi \rangle$$
holds for all $\beta \in TM_p^*$.

Remark 0.1.4 Let $\pi : E \to M$ be a smooth vector bundle and ∇ a connection on E. Then ∇ induces connections on the associate bundles of E (for instance, on the dual bundle or any tensor bundle).
For example, let $\pi : TM \to M$ be the tangent bundle and ∇ a connection on TM. Then there is precisely one connection ∇^* on TM^* so that for all $X, Y \in \mathfrak{X}(M)$ and $\alpha \in \mathcal{E}^{(1)}(M)$
$$\langle \nabla_X^* \alpha, Y \rangle = X \langle \alpha, Y \rangle - \langle \alpha, \nabla_X Y \rangle$$
holds, where $\langle \_,\_ \rangle$ is the natural pairing.
If $g \in \Gamma(M; T^{(0,2)}M)$, the connection induced on $T^{(0,2)}M$ is defined by
$$\nabla_\xi g(\eta, \zeta) = \xi(g(Y, Z)) - g(\nabla_\xi Y, \zeta) - g(\eta, \nabla_\xi Z),$$
with $\xi, \eta, \zeta \in TM_p$ and Y, Z continuations of η and ζ respectively.
If induced connections appear in the following discussions we will usually denote them again by ∇.

0.2 Concepts of curvature

In this part let (M, g) always be a semi-Riemannian manifold. We now introduce the curvature tensor, the Riemannian curvature tensor and their symmetries. We will not demonstrate these symmetry properties. But you can find the proofs in standard books about Riemannian geometry, see for example [37], §2.2.

Definition 0.2.1 Let $\pi : E \to M$ be a vector bundle, ∇ a connection on E. For $X, Y \in \mathfrak{X}(M), s \in \Gamma(M; E)$ we define
$$R : \mathfrak{X}(M) \times \mathfrak{X}(M) \times \Gamma(M; E) \to \Gamma(M; E) : (X, Y, s) \mapsto R(X, Y)s$$
$$R(X, Y)s := \nabla_X \nabla_Y s - \nabla_Y \nabla_X s - \nabla_{[X,Y]} s$$

and call R the *curvature of* ∇.

The value of $R(X,Y)s$ in $p \in M$ only depends on $X_p, Y_p \in TM_p$ and $s(p) \in E_p$. Therefore, it is possible to define the curvature R_p in p as a tri-linear map:

$$R_p : TM_p \times TM_p \times E_p \to E_p.$$

We thus can talk about the *curvature tensor* $R = (R_p)$. It is a global section in the vector bundle $TM^* \otimes TM^* \otimes E^* \otimes E$.

Remark 0.2.2 *a) The curvature tensor R ist anti-symmetric in X and Y: if we consider R to be a bilinear map from $\mathfrak{X}(M) \times \mathfrak{X}(M) \to End(\Gamma(M;E))$,*

$$(X,Y) \mapsto (R(X,Y) : s \mapsto R(X,Y)s),$$

the following holds:
$$R(Y,X) = -R(X,Y).$$

b) Let $\pi : TM \to M$ be the tangent bundle of M. If ∇ is a symmetric connection on M, its curvature tensor $R : \mathfrak{X}(M) \times \mathfrak{X}(M) \times \mathfrak{X}(M) \to \mathfrak{X}(M)$ satisfies the first Bianchi-identity:

$$R(X,Y)Z + R(Y,Z)X + R(Z,X)Y = 0,$$

for all $X, Y, Z \in \mathfrak{X}(M)$.

Definition 0.2.3 Let $\pi : E \to M$ be a vector bundle, ∇ a connection on E. If $R \in \Gamma(M; TM^* \otimes TM^* \otimes E^* \otimes E)$ is the curvature of ∇, we define $\mathrm{Rm} \in \Gamma(M; TM^* \otimes TM^* \otimes E^* \otimes E^*)$ by:

$$\mathrm{Rm}_p(\xi_1, \xi_2, e_1, e_2) := g_p(R_p(\xi_1, \xi_2)e_1, e_2).$$

We call Rm the *Riemannian curvature tensor* of ∇.

Remark 0.2.4 *a) Let $\pi : E \to M$ be a vector bundle and ∇ a metric connection. Then the Riemannian curvature tensor is anti-symmetric in the first two and in the last two arguments:*

$$\mathrm{Rm}(Y, X, s, t) = -\mathrm{Rm}(X, Y, s, t)$$
$$\mathrm{Rm}(X, Y, t, s) = -\mathrm{Rm}(X, Y, s, t),$$

for all $X, Y \in \mathfrak{X}(M)$ and $s, t \in \Gamma(M; E)$.

b) Let $\pi : TM \to M$ be the tangent bundle of M. If ∇ is a symmetric and metric connection on M, its Riemannian curvature tensor $\mathrm{Rm}_p : TM_p \times TM_p \times TM_p \times TM_p \to \mathbb{R}$ satisfies the first Bianchi-identity:

$$\mathrm{Rm}_p(\xi_1, \xi_2, \xi_3, \xi_4) + \mathrm{Rm}_p(\xi_2, \xi_3, \xi_1, \xi_4) + \mathrm{Rm}_p(\xi_3, \xi_1, \xi_2, \xi_4) = 0,$$

for all $\xi_1, \xi_2, \xi_3, \xi_4 \in TM_p$. Note that the first Bianchi-identity holds for every choice of three of the four arguments of the Riemannian curvature tensor.

c) Let $\pi : TM \to M$ again be the tangent bundle of M and ∇ a symmetric and metric connection on M.
Then the Riemannian curvature tensor $\mathrm{Rm}_p : TM_p \times TM_p \times TM_p \times TM_p \to \mathbb{R}$ fulfils the following symmetry condition:

$$\mathrm{Rm}_p(\xi_1, \xi_2, \xi_3, \xi_4) = \mathrm{Rm}_p(\xi_3, \xi_4, \xi_1, \xi_2),$$

for all $\xi_1, \xi_2, \xi_3, \xi_4 \in TM_p$.

Remark 0.2.5 *Let $\pi : TM \to M$ be the tangent bundle of M and ∇ a symmetric and metric connection on M.*

a) *If we now have a look at the covariant derivations $\nabla R \in \Gamma(M; TM^{(1,4)})$ and $\nabla \mathrm{Rm} \in \Gamma(M; TM^{(0,5)})$, we note that $\nabla_\xi R \in TM_p^{(1,3)}$ and $\nabla_\xi \mathrm{Rm} \in TM_p^{(0,4)}$ keep the same identities as R_p and Rm_p. For instance, the following holds:*

   (i) $\nabla_\xi R(\eta_2, \eta_1) = -\nabla_\xi R(\eta_1, \eta_2)$
   (ii) $\nabla_\xi R(\eta_1, \eta_2)\eta_3 + \nabla_\xi R(\eta_2, \eta_3)\eta_1 + \nabla_\xi R(\eta_3, \eta_1)\eta_2 = 0$
   (iii) $\nabla_\xi \mathrm{Rm}(\eta_1, \eta_2, \eta_4, \eta_3) = -\nabla_\xi \mathrm{Rm}(\eta_1, \eta_2, \eta_3, \eta_4)$

for all $\xi, \eta_1, \eta_2, \eta_3, \eta_4 \in TM_p$.

b) *Furthermore, the second Bianchi-identity holds:*

$$\nabla_\xi R(\eta, \zeta) + \nabla_\eta R(\zeta, \xi) + \nabla_\zeta R(\xi, \eta) = 0$$

for all $\xi, \eta, \zeta \in TM_p$.

Remark 0.2.6 If $g \in \Gamma(M; T^{(0,2)}M)$ is just symmetric (and possibly degenerate) and ∇ a symmetric and metric connection (with respect to g), the symmetries mentioned in (0.2.4) and (0.2.5) for the Riemannian curvature tensor also hold. This can easily be seen if you have a look at the proofs of these assertions, see for instance [37], §2.2. There you only have to use the symmetric and metric property of ∇ and it does not matter if g is degenerate or not.

Definition 0.2.7   a) The *Ricci tensor* $\mathrm{Ric} \in \Gamma(T^{(0,2)}M)$ is the contraction of the Riemannian curvature tensor in the first and fourth argument,

$$\mathrm{Ric} = C^{(1,4)}(\mathrm{Rm}).$$

b) The *scalar curvature* $S \in \Gamma(T^{(0,0)}M) = \mathcal{E}(M)$ is the contraction of the Ricci tensor,

$$S = C^{(1,2)}(\mathrm{Ric}) = \mathrm{tr}(\mathrm{Ric}).$$

0.2. Concepts of curvature

Remark 0.2.8 *The Ricci tensor* $\mathrm{Ric} \in \Gamma(M; T^{(0,2)}M)$ *is symmetric:*

$$\mathrm{Ric}_p(\xi, \eta) = \mathrm{Ric}_p(\eta, \xi),$$

for all $p \in M$ *and* $\xi, \eta \in TM_p$.

Notation 0.2.9 Let $p \in M$, U a neighbourhood of p and $x : U \to V \subseteq \mathbb{R}^n$ a chart. Then there are functions $g_{ij}, \Gamma^k_{ij}, R^l_{ijk}, \mathrm{Ric}_{ij} \in \mathcal{E}(V)$ so that

$$g|U = g_{ij} dx^i \otimes dx^j$$

$$\nabla_{\frac{\partial}{\partial x^i}} \frac{\partial}{\partial x^j} = \Gamma^k_{ij} \frac{\partial}{\partial x^k}$$

$$R|U = R^l_{ijk} dx^i \otimes dx^j \otimes dx^k \otimes \frac{\partial}{\partial x^l}$$

$$\mathrm{Ric}|U = \mathrm{Ric}_{ij} dx^i \otimes dx^j.$$

It follows that the *Christoffel symbols* are given by

$$\Gamma^r_{ij} = \frac{1}{2} g^{kr} (\partial_i g_{jk} + \partial_j g_{ik} - \partial_k g_{ij}),$$

the components of the Riemannian curvature tensor by

$$R^s_{ijk} = \partial_j \Gamma^s_{ki} - \partial_k \Gamma^s_{ji} + \Gamma^s_{jr} \Gamma^r_{ki} - \Gamma^s_{kr} \Gamma^r_{ij},$$

and those of the Ricci tensor by

$$\mathrm{Ric}_{ik} = R^j_{ijk} = \partial_j \Gamma^j_{ki} - \partial_k \Gamma^j_{ji} + \Gamma^j_{jr} \Gamma^r_{ki} - \Gamma^j_{kr} \Gamma^r_{ji}.$$

Example 0.2.10 We give an example which is interesting for our case. Let M be a smooth manifold and $g \in \Gamma(M; T^{(0,2)}M)$. Let furthermore ∇ be a connection on the tangent bundle of M. We discuss the condition

$$\nabla g = 0$$

in detail. Therefore:
Let $p \in M$, U a neighbourhood of p, $x : U \to V \subseteq \mathbb{R}^n$ a chart and $g|U = g_{ij} dx^i \otimes dx^j$ with $g_{ij} \in \mathcal{E}(V)$. For $g \in \Gamma(M; T^{(0,2)}M)$, $\nabla g \in \Gamma(M; T^{(0,3)}M)$ and so

$$\nabla g|U =: \nabla_i g_{jk} dx^i \otimes dx^j \otimes dx^k,$$

thus

$$\nabla_{\frac{\partial}{\partial x^i}} g = \nabla_i g_{jk} dx^j \otimes dx^k.$$

In order to calculate $\nabla_i g_{jk}$, we have to consider the following:

$$\begin{aligned}
\nabla_{\frac{\partial}{\partial x^i}} g &= \frac{\partial}{\partial x^i} g_{jk} dx^j \otimes dx^k + g_{jk} \nabla_{\frac{\partial}{\partial x^i}} dx^j \otimes dx^k + g_{jk} dx^j \otimes \nabla_{\frac{\partial}{\partial x^i}} dx^k \\
&= D_i g_{jk} dx^j \otimes dx^k + g_{lk} \nabla_{\frac{\partial}{\partial x^i}} dx^l \otimes dx^k + g_{jl} dx^j \otimes \nabla_{\frac{\partial}{\partial x^i}} dx^l \\
&= D_i g_{jk} dx^j \otimes dx^k + g_{lk}(-\Gamma^l_{ij}) dx^j \otimes dx^k + g_{jl}(-\Gamma^l_{ik}) dx^j \otimes dx^k \\
&= (D_i g_{jk} - \Gamma^l_{ij} g_{lk} - \Gamma^l_{ik} g_{jl}) dx^j \otimes dx^k \\
&= \nabla_i g_{jk} dx^j \otimes dx^k.
\end{aligned}$$

If $\nabla g = 0$, then

$$\nabla_i g_{jk} = D_i g_{jk} - \Gamma^l_{ij} g_{kl} - \Gamma^l_{ik} g_{jl} = 0,$$

for all $i, j, k \in \{1, ..., n\}$ and for all charts.

Definition 0.2.11 Let $T \in \Gamma(M; T^{(0,2)}M)$ be a symmetric $(0,2)$-tensor field. We then define the divergence of T, $\mathrm{div}(T) \in \Gamma(M; T^{(0,1)}M) = \mathcal{E}^{(1)}(M)$ as

$$\mathrm{div}(T) = \mathrm{tr}_{(1,2)}(\nabla T),$$

which means

$$\langle \mathrm{div}(T)(p), \xi \rangle = \sum_{i=1}^n \nabla_{e_i} T(e_i, \xi),$$

for all $p \in M$, $\xi \in TM_p$ and (e_i) an orthonormal basis of TM_p.

Proposition 0.2.12 If $\mathrm{Ric} \in \Gamma(M; T^{(0,2)}M)$ is the Ricci tensor and $S \in \mathcal{E}(M)$ the scalar curvature, the following holds (known as the contracted second Bianchi-identity):

$$\mathrm{div}(\mathrm{Ric}) = \frac{1}{2} dS.$$

(Again the proof can be found in several standard books about Riemannian geometry, see for instance [37], §2.2.)

Definition 0.2.13 The Einstein curvature tensor $\mathrm{Ein}(g)$ of (M, g) is given by

$$\mathrm{Ein}(g) = \mathrm{Ric}(g) - \frac{S}{2} g.$$

Remark 0.2.14 It also holds: $\mathrm{div}(Sg) = dS$. Therefore, the divergence of the Einstein tensor disappears, which can be seen by:

$$\mathrm{Ein}(g) = \mathrm{Ric}(g) - \frac{1}{2} Sg$$
$$\Rightarrow \mathrm{div}(\mathrm{Ein}(g)) = \mathrm{div}\left(\mathrm{Ric}(g) - \frac{1}{2} Sg\right) = \mathrm{div}(\mathrm{Ric}(g)) - \frac{1}{2} \mathrm{div}(Sg) = \frac{1}{2} dS - \frac{1}{2} dS = 0.$$

0.3 Newton's theory of gravitation and Einstein's theory of relativity

In this part we want to introduce the basic concepts of the two theories. But first we have to cite some definitions and theorems from Linear Algebra which are also important for the following chapters. For the proofs of the theorems and further information see [17], §5.7, for instance.

Definition 0.3.1 Let V be a finite-dimensional real vector space and V^* the dual space. The annihilator of $v \in V$, $v \neq 0$, is defined as

$$\text{Ann}(v) := \{\lambda \in V^* : \lambda(v) = 0\}.$$

Definition 0.3.2 Let V be a finite-dimensional real vector space and $g : V \times V \to \mathbb{R}$ a symmetric bilinear form. We call g

- *positive definite*, if $g(v,v) > 0$ for all $v \in V \setminus \{0\}$.
- *positive semi-definite*, if $g(v,v) \geq 0$ for all $v \in V$.
- *negative definite*, if $g(v,v) < 0$ for all $v \in V \setminus \{0\}$.
- *negative semi-definite*, if $g(v,v) \leq 0$ for all $v \in V$.
- *indefinite*, otherwise.

Remark 0.3.3 Now let V be a four-dimensional, real vector space and $v \in V$ with $v \neq 0$. Due to the dimension formula we then have $\dim(\text{Ann}(v)) = 3$ because of the fact that $\text{Ann}(v) = \ker(f)$, where $f : V^* \to \mathbb{R}$, $f(\lambda) = \lambda(v)$ and $f \neq 0$.

Now let $g : V \times V \to \mathbb{R}$ be a symmetric bilinear form. Then there are subspaces $V_+, V_- \subseteq V$ so that

$$V = V_0 \oplus V_- \oplus V_+$$

where $V_0 = \{v \in V : g(v,w) = 0, \forall w \in V\}$ and

$$g|_{V_- \times V_-} \text{ negative definite}$$

and

$$g|_{V_+ \times V_+} \text{ positive definite}.$$

Definition 0.3.4 We call the symmetric bilinear form $g : V \times V \to \mathbb{R}$ *non-degenerate*, if

$$\text{Deg}(g) := V_0 = (0).$$

Remark 0.3.5 If we consider the transition from the tangent to the cotangent space and vice versa defined in (0.1.2),

$$\varphi : V \to V^*, \ \langle \varphi(v), w \rangle = g(v, w),$$

g is non-degenerate if and only if φ is an isomorphism.

Theorem (of Sylvester) 0.3.6 *Let V be a finite-dimensional real vector space, $g : V \times V \to \mathbb{R}$ a symmetric bilinear form and*

$$V = V_0 \oplus V_- \oplus V_+,$$

so that V_0, V_- and V_+ are defined as above. Then the numbers $n_0 := \dim V_0$, $n_- := \dim V_-$ and $n_+ := \dim V_+$ are unambiguously defined, which means that they do not depend on the choice of V_- and V_+.
Moreover:

$$n_- = \max\{\dim W : W \subseteq V \text{ subspace and } s(v,v) < 0 \text{ for } v \in W, \ v \neq 0\},$$
$$n_+ = \max\{\dim W : W \subseteq V \text{ subspace and } s(v,v) > 0 \text{ for } v \in W, \ v \neq 0\}.$$

Due to theorem (0.3.6) it now makes sense to formulate the following

Definition 0.3.7 We define the *index of g* by $\mathrm{ind}(g) := n_-$ as well as the *rank of g* by $\mathrm{rk}(g) := n_- + n_+$.

It can immediately be seen:

Lemma 0.3.8 *Let V be a finite-dimensional real vector space, $n = \dim V$ and $g : V \times V \to \mathbb{R}$ a symmetric bilinear form. Then, g is*

- *positive definite, if and only if $\mathrm{rk}(g) = n$ and $\mathrm{ind}(g) = 0$.*
- *positive semi-definite, if and only if $\mathrm{rk}(g) \leq n$ and $\mathrm{ind}(g) = 0$.*
- *negative definite, if and only if $\mathrm{rk}(g) = n$ and $\mathrm{ind}(g) = n$.*
- *negative semi-definite, if and only if $\mathrm{rk}(g) = p \leq n$ and $\mathrm{ind}(g) = p$.*
- *indefinite, if and only if $0 < \mathrm{ind}(g) < \mathrm{rk}(g)$.*

Corollary 0.3.9 *Let V be a finite-dimensional real vector space, $g : V \times V \to \mathbb{R}$ a symmetric bilinear form and n_- as well as n_+ its invariants. Then we can find a basis $\mathfrak{B}$ of V so that*

$$\mathcal{M}_{\mathfrak{B}}(g) = \begin{pmatrix} E_{n_+} & 0 & 0 \\ 0 & -E_{n_-} & 0 \\ 0 & 0 & 0 \end{pmatrix},$$

where $E_n \in \text{Mat}(n, \mathbb{R})$ is the unit matrix.

In particular, it is possible to find an orthogonal decomposition, which means that we can find V_- and V_+ so that
$$V = V_0 \oplus V_- \oplus V_+.$$

Remark 0.3.10 Let V be a real vector space, $g : V \times V \to \mathbb{R}$ a bilinear form which is positive semi-definite. If $g(v,v) = 0$, then $v \in \text{Deg}(g) := V_0$.

Proof. According to (0.3.9) we can find an orthogonal decomposition of V of the form
$$V = V_0 \oplus V_+.$$

Now let $v \in V$ be arbitrary with $g(v,v) = 0$. Then there is a $v_0 \in V_0$ and a $v_+ \in V_+$ so that $v = v_0 + v_+$. Now we have
$$0 = g(v,v) = \underbrace{g(v_0, v_0)}_{=0,\ v_0 \in V_0} + 2\underbrace{g(v_0, v_+)}_{=0,\ V_0 \perp V_+} + g(v_+, v_+) = g(v_+, v_+).$$

But this means that $v_+ = 0$ and therefore $v \in \text{Deg}(g)$. □

Definition 0.3.11 a) We call a semi-Riemannian manifold (M, g) a *Lorentzian manifold*, if g has index 1 (which means that $g_p : TM_p \times TM_p \to \mathbb{R}$ has index 1 for every $p \in M$). We then call g a *Lorentzian metric*.

b) Let (M, g) be a four-dimensional, connected Lorentzian manifold. Then
$$\mathcal{T}_p = \{\xi \in TM_p : g_p(\xi, \xi) < 0\}$$

is the set of all *timelike vectors*,
$$\Lambda_p = \{\xi \in TM_p : g_p(\xi, \xi) = 0\} \setminus \{0\}$$

is the set of all *null vectors*,
$$\mathcal{U}_p = \{\xi \in TM : g_p(\xi, \xi) > 0\} \cup \{0\}$$

is the set of all *spacelike vectors*.

c) A four-dimensional, connected Lorentzian manifold (M, g) is called *time-oriented*, if there is a timelike smooth vector field $X \in \mathfrak{X}(M)$.
$\mathcal{T}_p$ consists of two connected components σ_p and τ_p, where $\tau_p = -\sigma_p$. Two timelike vector fields X and Y are considered to be *equivalent*, if X_p and Y_p belong to the same connected component of $\mathcal{T}_p$, for all $p \in M$. An equivalence class $[X]$ of timelike vector fields is called *time-orientation* on (M, g).

If (M, g) is time-oriented there are exactly two time-orientations. If we choose one with

0.3. Newton's theory of gravitation and Einstein's theory of relativity

$[X]$, we call a timelike tangent vector $\xi \in TM_p$ *future-oriented*, if X_p and ξ belong to the same component. Otherwise $\xi \in TM_p$ is called *past-oriented*.

d) A four-dimensional connected and time-oriented Lorentzian manifold $(M, g, [X])$ is called *spacetime*.

Now we have a look at Newton's theory of gravitation. We therefore first have to explain the concept of a Newtonian spacetime.

Definition 0.3.12 a) The *Newtonian space* is a three-dimensional euclidian space E, thus E is isometrical to the Riemannian manifold $(\mathbb{R}^3, g_{\text{eucl.}})$.

b) The *Newtonian time* is represented by the real axis $\mathbb{R}$.

c) The Riemannian product manifold $(\mathbb{R}, g_{\text{eucl.}}) \times E$ of Newtonian time and Newtonian space is called *Newtonian spacetime*.

Now we want to discuss the Newtonian law of motion and Newton's law of gravitation:

Law of motion 0.3.13 Let $x : I \to \mathbb{R} \times \mathbb{R}^3$ be a curve, called *Newtonian particle* of mass m which is subject to the *force* $F : \mathbb{R} \times \mathbb{R}^3 \to \mathbb{R}^3$. Then the following holds:

$$\frac{d(m\dot{x})}{dt} = m\ddot{x} = F.$$

Definition 0.3.14 Let $K \to M(K)$ be a mass distribution. A map $\rho : \mathbb{R} \times \mathbb{R}^3 \to \mathbb{R}_+$ is called *mass density for M*, if for every $K \subseteq \mathbb{R} \times \mathbb{R}^3$ compact

$$\int_K \rho(x)dx =: M(K)$$

holds.

Law of gravitation 0.3.15 Let $\rho : \mathbb{R} \times \mathbb{R}^3 \to \mathbb{R}_+$ be a mass density for M and $u : \mathbb{R} \times \mathbb{R}^3 \to \mathbb{R}$ a map. We say that u satisfies the *Newtonian law of gravitation* if

$$\Delta u = \frac{\partial^2 u}{(\partial x^1)^2} + \frac{\partial^2 u}{(\partial x^2)^2} + \frac{\partial^2 u}{(\partial x^3)^2} = 4\pi\rho.$$

We then call $G : \mathbb{R} \times \mathbb{R}^3 \to \mathbb{R}^3$ with

$$G = -\operatorname{grad}(u),$$

the *gravitation field* for u.

Example: A perfect fluid 0.3.16 In continuum mechanics, a *perfect fluid* is described by a *velocity field* $V : \mathbb{R} \times \mathbb{R}^3 \to \mathbb{R}^3$, a *pressure function* $p : \mathbb{R} \times \mathbb{R}^3 \to \mathbb{R}$ and a *mass density* $\rho : \mathbb{R} \times \mathbb{R}^3 \to \mathbb{R}$ which is constant (but not zero) for a perfect fluid. The integral curve of the

0.3. Newton's theory of gravitation and Einstein's theory of relativity

velocity field describes the mean movement of the fluid molecules, while the other functions describe the energy of the fluid molecules per volume and its internal interactions.

Then, the *Euler equations*, which express the preservation of energy and momentum, describe the relation between (V, ρ, p) and the gravitation field $G = -\operatorname{grad}(u)$ which they cause. The following holds:

$$\rho D_t V = -\operatorname{grad} p + \rho G$$
$$\operatorname{div} V = 0.$$

If you are interested in further reading see, for instance, [21], §4.2.

If we now talk about Einstein's theory of gravitation we first have to note that here gravitation is not represented by a vector field but by the curvature of the space which now is represented by a spacetime (M, g).

Now let $x : [0, 1] \to M$ be a curve which satisfies *Einstein's law of motion*

$$\nabla_t \dot{x} = 0.$$

Then x is called a *free falling particle* as it is not subject to a force anymore. The law of motion thus demands that x is a geodesic on M.

If we now discuss the law of gravitation, we note that the present mass is represented by a symmetric *energy momentum tensor* $T \in \Gamma(M, T^{(0,2)}M)$.

Einstein's law of gravitation, known as *Einstein's field equations*, is then given by

Einstein's field equations 0.3.17 Let (M, g) be a spacetime. The metric g has to satisfy the condition

$$\operatorname{Ein}(g) = 8\pi T,$$

where $\operatorname{Ein}(g)$ is the Einstein curvature tensor and $T \in \Gamma(M; T^{(0,2)}M)$ an energy momentum tensor.

Remark 0.3.18 If $T = 0$, M is called a *vacuum*. Then for the field equations the following holds:

$$0 = \operatorname{tr}(\operatorname{Ein}(g)) = \operatorname{tr}_g(\operatorname{Ric}) - \frac{S}{2}\operatorname{tr}_g g = S - \frac{S}{2}n = (1 - \frac{n}{2})S.$$

This means that $S = 0$ for $\dim M = n \neq 2$ and it follows from the definition of the Einstein curvature tensor that $\operatorname{Ein}(g) = \operatorname{Ric}(g)$. Therefore,

$$\operatorname{Ric}(g) = 0$$

is called *Einstein vacuum equation*.

Chapter 1

Frame theory

In this section we start to discuss the frame theory which goes back to Jürgen Ehlers. His frame theory makes it possible to compare models of General Relativity with those of Newton's theory of gravitation. Although the fundamental concepts come from Jürgen Ehlers this chapter is not a repetition of his former work or the work of some of his students (for example Lottermoser, [31]). We here try to adopt a new approach and try to avoid coordinate calculations as long as possible. Furthermore, we use results from Algebraic Topology in order to have a better global understanding of the implications of the frame theory.

1.1 The structure of the frame theory

The structure of Ehlers' frame theory is determined by the following ingredients, axioms and definitions.

Ingredients 1.1.1 If you want to establish a frame theory which contains both Newton's and Einstein's theories of gravitation you can use the following mathematical objects:

- a four-dimensional connected smooth manifold M (which is Hausdorff by definition of manifolds);
- a *causality constant* $\lambda \in \mathbb{R}$;
- a symmetric tensor field g on the tangent bundle of M, which means $g \in \Gamma(M; T^{(0,2)}M)$, called *time metric*;
- a symmetric tensor field h on the cotangent bundle of M, which means $h \in \Gamma(M; T^{(2,0)}M)$, called *space metric*;
- a symmetric tensor field T on the cotangent bundle of M, $T \in \Gamma(M; T^{(2,0)}M)$, called *matter tensor*;
- a so called *gravity field* ∇, a symmetric connection on M.

1.1. The structure of the frame theory

Axioms 1.1.2 The introduced objects λ, g, h, T and ∇ have to satisfy the following axioms:

1. At every point $p \in M$ there is a *timelike tangent vector* $\xi \in TM_p$, which means:

$$g_p(\xi, \xi) > 0,$$

and $h_p : TM_p^* \times TM_p^* \to \mathbb{R}$ is positive definite on the annihilator of ξ,

$$h_p|_{\text{Ann}(\xi) \times \text{Ann}(\xi)} > 0,$$

where $\text{Ann}(\xi) = \{\mu \in TM_p^* : \mu(\xi) = 0\}$.

2. For the transitions from the tangent space to the cotangent space and vice versa induced by g and h (see (0.1.2)), it is demanded that

$$\varphi_p \circ \psi_p = -\lambda \, \text{id}_{TM_p^*}$$

for all $p \in M$.

3. Both g and h have to be parallel with regard to the connection ∇, which means

$$\nabla g = \nabla h = 0,$$

where ∇ here denotes connections on $TM \otimes TM$ and $TM^* \otimes TM^*$ induced by ∇ (see also (0.1.4)).

4. The matter tensor T has to satisfy $\text{div}(T) = 0$.

5. The curvature tensor R of the connection ∇ induces a tensor $R^\#$ by

$$R_p^\#(\xi, \alpha, \eta, \beta) := \langle \beta, R_p(\xi, \psi_p(\alpha))\eta \rangle,$$

for $\xi, \eta \in TM_p$ and $\alpha, \beta \in TM_p^*$. This tensor $R^\#$ is a four-linear map on $TM_p \times TM_p^* \times TM_p \times TM_p^*$ for every $p \in M$, and we require that it is symmetric in the two first and the two last arguments,

$$R_p^\#(\xi, \alpha, \eta, \beta) = R_p^\#(\eta, \beta, \xi, \alpha),$$

for all $\xi, \eta \in TM_p$ and $\alpha, \beta \in TM_p^*$.

6. Finally, the frame theory demands:

$$\text{Ric}(\nabla) = 8\pi \left(T^{bb} - \frac{1}{2} \text{tr}(T^b) g \right).$$

Remark 1.1.3   a) Let us first have a look at the second axiom. In case of $\lambda \neq 0$ it follows that φ is surjective and ψ injective, and so both are bijective due to $\dim(TM_p) = \dim(TM_p^*) < \infty$. Therefore, g and h are non-degenerate and we have: $\psi = -\lambda \varphi^{-1}$.

b) We have to comment on the fourth axiom. As already mentioned in (0.2.11), divergence usually is defined for a symmetric $(0,2)$–tensor field T. In our case we now have a $(2,0)$–tensor field T. But as ∇T then is a $(2,1)$–tensor field, $\nabla T \in T^{(2,1)}M$, we now define

$$\mathrm{div}(T) = \mathrm{tr}_{(1,3)}(\nabla T),$$

which means that

$$\langle \beta, \mathrm{div}(T)(p) \rangle = \sum_{i=1}^{n} \nabla_{e_i} T(\alpha^i, \beta),$$

where (e_i) is an orthonormal basis of TM_p and (α^i) the dual basis. Note that $\mathrm{div}(T) \in T^{(1,0)}M$. As T is symmetric, it does not matter if we build the $(1,3)-$ or the $(2,3)-$trace.

c) Now let us deal with the fifth axiom and the definition of $R_p^\#$. Note that in case of $\lambda \neq 0$ we rediscover the definition of the Riemannian curvature tensor (consider also d)). But the fifth axiom can not be replaced by a corresponding formula of the Riemannian curvature tensor as this is not possible in case of $\lambda = 0$.

d) In case of $\lambda \neq 0$, (M, g) is a pseudo-Riemannian manifold. Then (h, ∇, T) are determined by g: h is the quasi-inverse of g (see axiom two), T is defined by axiom six and condition three determines the connection ∇ definitely. According to the precondition the connection ∇ is symmetric and so it is the Levi-Civita connection of g. This is not the case for $\lambda = 0$, of course.

It can also be shown that some conditions can be neglected in case of $\lambda \neq 0$. As ∇ is the Levi-Civita connection of g, the curvature tensor and the Riemannian curvature tensor satisfy the usual symmetries (see also (0.2.4) and (0.2.2)). For instance, condition number five follows directly from the metric condition three and the fact that the connection is symmetric, since:

Let $\xi, \eta \in TM_p$, as well as $\alpha, \beta \in TM_p^*$. Then:

$$\begin{aligned}
R_p^\#(\xi, \alpha, \eta, \beta) &= \langle \beta, R_p(\xi, \psi_p(\alpha))\eta \rangle = -\frac{1}{\lambda}\langle \varphi_p(\psi_p(\beta)), R_p(\xi, \psi_p(\alpha))\eta \rangle \\
&= -\frac{1}{\lambda} g_p(\psi_p(\beta), R_p(\xi, \psi_p(\alpha))\eta) = -\frac{1}{\lambda} g_p(R_p(\xi, \psi_p(\alpha))\eta, \psi_p(\beta)) \\
&= -\frac{1}{\lambda} Rm_p(\xi, \psi_p(\alpha), \eta, \psi_p(\beta)) = -\frac{1}{\lambda} Rm_p(\eta, \psi_p(\beta), \xi, \psi_p(\alpha)) \\
&= -\frac{1}{\lambda} g_p(R_p(\eta, \psi_p(\beta))\xi, \psi_p(\alpha)) = -\frac{1}{\lambda}\langle \varphi_p(\psi_p(\alpha)), R_p(\eta, \psi_p(\beta))\xi \rangle \\
&= \langle \alpha, R_p(\eta, \psi_p(\beta))\xi \rangle = R_p^\#(\eta, \beta, \xi, \alpha),
\end{aligned}$$

where $\langle \_, \_ \rangle$ again denotes the natural pairing.

e) Moreover, we have to comment on the last axiom. Usually the reader should expect Einstein's field equations. However, note that these are equivalent to axiom six in case of $\lambda \neq 0$ which is demonstrated by the following:

It is known that Einstein's field equations are given by:

$$\mathrm{Ein}(g) = 8\pi \widetilde{T},$$

where $\mathrm{Ein}(g) = \mathrm{Ric}(\nabla) - \frac{1}{2}Sg$ and $\widetilde{T} \in \Gamma(M; T^{(0,2)}M)$. If you now build the trace of the equation $\mathrm{Ric}(\nabla) - \frac{1}{2}Sg = 8\pi\widetilde{T}$ with regard to g, you get

$$S - \frac{1}{2}S \cdot 4 = 8\pi \, \mathrm{tr}_g(\widetilde{T}),$$

thus

$$S = -8\pi \, \mathrm{tr}_g(\widetilde{T}).$$

If you put this in $\mathrm{Ric}(\nabla) - \frac{1}{2}Sg = 8\pi\widetilde{T}$, you get

$$\mathrm{Ric}(\nabla) = 8\pi\widetilde{T} + \frac{1}{2}Sg = 8\pi\widetilde{T} - 4\pi \, \mathrm{tr}_g(\widetilde{T})g$$

$$= 8\pi \left(\widetilde{T} - \frac{1}{2}\mathrm{tr}_g(\widetilde{T})g\right) = 8\pi \left(T^{bb} - \frac{1}{2}\mathrm{tr}(T^b)\right),$$

with $\widetilde{T} \in \Gamma(M; T^{(0,2)}M)$, thus $T = \widetilde{T}^{\#\#}$ and so $\widetilde{T} = T^{bb}$ and $\widetilde{T}^{\#} = T^b$. As you do not lose any information by the transition $T \to \widetilde{T} = T^{bb}$ in case of $\lambda \neq 0$ these equations are equivalent to Einstein's field equations. This is clearly not the case for $\lambda = 0$.

f) Now you can see that condition four follows immediately from condition six in case of $\lambda \neq 0$. From (0.2.14) follows that

$$\mathrm{div}(\mathrm{Ein}) = 0$$

and therefore also

$$\mathrm{div}(T^{bb}) = 0.$$

g) It is really surprising that the frame theory does not determine a time orientation, which usually belongs to every model in General Relativity. Normally one would expect that an equivalence class $[B]$ is mentioned, where $B \in \mathfrak{X}(M)$ is a timelike global vector field (see 0.3.11).

h) If you have a look at the axioms mentioned by Ehlers in his papers (see for instance [13]), our first axiom there is divided into the following two axioms:

1. At every point $p \in M$ there is at least one timelike tangent vector $\xi \in TM_p$.
2. For every $\xi \in TM_p$ with $g_p(\xi, \xi) > 0$,

$$h_p|_{\mathrm{Ann}(\xi) \times \mathrm{Ann}(\xi)} > 0.$$

We will use our axiom one in the following sections and we will see there that although the formulations of Ehlers seem to be stronger, our axiom number one is equivalent to the two of Ehlers (for $\lambda = 0$ and this is the only case where we need this axiom). We therefore are allowed to exchange the axioms (see (1.2.6)).

i) Sometimes a gravity constant $G \in \mathbb{R}_+$ also appears as one of the objects of the frame theory. But as there are no conditions for this gravity constant we consider G to be equal to 1, $G = 1$, and will not mention it any further.

After the presentation of the axioms of the frame theory we now analyse the situation of the frame theory in one point. We therefore have to discuss some concepts of Linear Algebra.

1.2 Linear Algebra

In order to understand how we can identify Newton's theory of gravitation and the theory of General Relativity in the frame theory, we are first interested in the rank and index of the bilinear forms g and h in dependence on λ. (For the definitions and the Sylvester's theorem see the beginning of section (0.3).) In the following discussion let V be a real vector space of dimension 4, $g : V \times V \to \mathbb{R}$ and $h : V^* \times V^* \to \mathbb{R}$ symmetric bilinear forms.

Theorem 1.2.1 *Let V, g and h be given as mentioned above, as well as $\lambda \in \mathbb{R}$. They are required to satisfy the following conditions:*

1. *There is a $v \in V$ with $g(v,v) > 0$ and $h|_{Ann(v) \times Ann(v)} > 0$.*

2. *For*
$$\varphi : V \to V^*, \ \langle \varphi(v), w \rangle = g(v,w)$$
$$\psi : V^* \to V, \ \langle \alpha, \psi(\beta) \rangle = h(\alpha, \beta)$$
with $v, w \in V$ and $\alpha, \beta \in V^*$:
$$\varphi \circ \psi = -\lambda \operatorname{id}_{V^*}.$$

Then:

(i) For $\lambda > 0$, g and h have rank 4, g has index 3 and h has index 1.

(ii) For $\lambda < 0$, g and h have rank 4 and both have index 0.

(iii) For $\lambda = 0$, g has rank 1 and h has rank 3. Both have index 0.

Proof. First let $\lambda \neq 0$. Then we know by the second condition that the maps
$$\varphi : V \to V^*, \ \langle \varphi(v), w \rangle = g(v,w)$$
$$\psi : V^* \to V, \ \langle \alpha, \psi(\beta) \rangle = h(\alpha, \beta)$$
are isomorphisms (remember that from $\varphi \circ \psi = -\lambda \operatorname{id}$ follows that φ is surjective and ψ injective and due to $\dim(TM_p) = \dim(TM_p^*) < \infty$ that φ and ψ are bijective) and it follows that g and

h are non-degenerate. But this means that $\text{rk}(g) = \text{rk}(h) = 4$.

Now we have a look at the first condition. It immediately shows that $\text{ind}(g) \leq 3$. Furthermore, there is a three-dimensional subspace of V^* on which h is positive definite. So we get $\text{ind}(h) \leq 1$.
Let $v \in V$ and $\alpha = \psi^{-1}(v)$, so $v = \psi(\alpha)$. Then:

$$g(v,v) = \langle \varphi(v), v \rangle = \langle \varphi \circ \psi(\alpha), \psi(\alpha) \rangle = -\lambda \langle \alpha, \psi(\alpha) \rangle = -\lambda h(\alpha, \alpha). \tag{1.1}$$

Now let $\lambda > 0$. Due to equation (1.1) we get

$$\text{ind}(g) = \text{ind}(-h) \Leftrightarrow \text{ind}(h) = \text{ind}(-g).$$

As $\text{ind}(g) \leq 3$, we have $\text{ind}(-g) \geq 1$. Due to $\text{ind}(h) \leq 1$ it follows that

$$\text{ind}(-g) = \text{ind}(h) = 1 \text{ and } \text{ind}(g) = 3,$$

which shows (i).

Now let $\lambda < 0$. Due to (1.1) we have

$$\text{ind}(g) = \text{ind}(h).$$

As $\text{ind}(h) \leq 1$ we just have to exclude that $\text{ind}(g) = \text{ind}(h) = 1$. Let us assume that this is the case. Then there is a decomposition $V = V_+ \oplus V_-$ with $g|_{V_+ \times V_+} > 0$, $g|_{V_- \times V_-} < 0$ and even $g(V_+, V_-) = 0$ (see (0.3.9)). If now $w \in V_- \setminus \{0\}$ (which exists as we assumed that $\dim V_- = 1$) and $\beta = \varphi(w)$, then

$$h(\beta, \beta) = \langle \beta, \psi(\beta) \rangle = \langle \beta, \psi \circ \varphi(w) \rangle = -\lambda \langle \varphi(w), w \rangle = -\lambda g(w, w) < 0.$$

But on the other hand, for $v \in V_+ \setminus \{0\}$

$$\langle \beta, v \rangle = \langle \varphi(w), v \rangle = g(v, w) = 0$$

holds, so $\beta \in \text{Ann}(v)$, which is a contradiction to condition 2. Therefore,

$$\text{ind}(g) = \text{ind}(h) = 0,$$

which shows (ii).

Now let $\lambda = 0$. Condition one provides that $\text{rk}(g) \geq 1$. Furthermore, it implies that $\text{rk}(h) \geq 3$ as h is positive definite on the annihilator of v, which is three-dimensional. From the second condition follows that $\text{rk}(g) + \text{rk}(h) \leq 4$ due to the following calculation:

$$\varphi \circ \psi = 0 \Rightarrow \text{im}(\psi) \subseteq \ker(\varphi) \Rightarrow \text{rk}(\psi) \leq \dim(\ker(\varphi)) = 4 - \text{rk}(\varphi).$$

This clearly shows that g has rank 1 and h has rank 3. Due to the first condition g therefore has index 0 as claimed. But as h has to be positive definite on a three-dimensional subspace of V^* the index of h also equals 0. This shows (iii) and thus the claim of the theorem. □

Now we will prove a normal form theorem for g and h. First let us have a look at this preliminary remark:

Lemma 1.2.2 *Let V, g and h be the same as before as well as $\lambda \in \mathbb{R}$ and let*

$$\varphi : V \to V^*, \; \langle \varphi(v), w \rangle = g(v, w)$$
$$\psi : V^* \to V, \; \langle \alpha, \psi(\beta) \rangle = h(\alpha, \beta)$$

be the transitions between V and V^*, where $v, w \in V$ and $\alpha, \beta \in V^*$. If these transitions satisfy the condition

$$\varphi \circ \psi = -\lambda \, \mathrm{id}_{V^*},$$

then also follows

$$\psi \circ \varphi = -\lambda \, \mathrm{id}_V .$$

Proof. First let $\lambda \neq 0$. Then the claim directly follows as φ and ψ are invertible and we have $\varphi = -\lambda \psi^{-1}$ as well as $\psi = -\lambda \varphi^{-1}$. So we get:

$$\psi \circ \varphi = -\lambda \varphi^{-1} \circ \varphi = -\lambda \, \mathrm{id}_V .$$

For $\lambda = 0$ we first define

$$H := \ker(\varphi) = \mathrm{Deg}(g) = \{v \in V : g(v, w) = 0 \; \forall w \in V\} \subseteq V.$$

As $\dim(\mathrm{im}(\varphi)) = \mathrm{rk}(g) = 1$, the dimension formula leads to

$$\dim(\ker(\varphi)) = 4 - \dim(\mathrm{im}(\varphi)) = 4 - 1 = 3.$$

Moreover, we have $\dim(\mathrm{im}(\psi)) = \mathrm{rk}(h) = 3$. Due to the fact that $\varphi \circ \psi = 0$ in case of $\lambda = 0$, it immediately follows that $\mathrm{im}(\psi) \subseteq \ker(\varphi)$ and on top of that, because of the dimensions, we have $\mathrm{im}(\psi) = \ker(\varphi) = H$. Now let $v \in V$ and $\alpha \in V^*$. Then

$$h(\alpha, \varphi(v)) = \langle \varphi(v), \psi(\alpha) \rangle = g(v, \underbrace{\psi(\alpha)}_{\in H}) = 0, \; \forall \alpha \in V^*,$$

and we get $\varphi(v) \in \mathrm{Deg}(h) = \ker(\psi)$, $\forall v \in V$. So we have $\mathrm{im}(\varphi) \subseteq \ker(\psi)$ and eventually $\psi \circ \varphi = 0$. □

Remark 1.2.3 If we have a look at the dimensions of the subspaces of V and V^* mentioned before we can even see that $\mathrm{im}(\varphi) = \ker(\psi)$ because of the fact that

$$\dim(\mathrm{im}(\varphi)) = \dim(\mathrm{rk}(g)) = 1 = 4 - 3 = 4 - \mathrm{rk}(h) = 4 - \dim(\mathrm{im}(\psi)) = \dim(\ker(\psi)).$$

Due to this we can define

$$H := \ker(\varphi) = \mathrm{Deg}(g) = \mathrm{im}(\psi) \subseteq V,$$
$$L := \ker(\psi) = \mathrm{Deg}(h) = \mathrm{im}(\varphi) \subseteq V^*.$$

Normal form theorem 1.2.4 *Let V, g and h be the same as above, as well as $\lambda \in \mathbb{R}$. They are required to satisfy the following conditions:*

1. *There is a $v \in V$ with $g(v,v) > 0$ and $h|_{Ann(v) \times Ann(v)} > 0$.*

2. *For*

$$\varphi : V \to V^*, \quad \langle \varphi(v), w \rangle = g(v,w)$$
$$\psi : V^* \to V, \quad \langle \alpha, \psi(\beta) \rangle = h(\alpha, \beta),$$

where $v, w \in V$ and $\alpha, \beta \in V^*$,

$$\varphi \circ \psi = -\lambda \, \mathrm{id}_{V^*}$$

holds (and as we have seen before this implies $\psi \circ \varphi = -\lambda \, \mathrm{id}_V$).

Then there is a basis $\mathfrak{A} = (e_0, ..., e_3)$ of V so that we have for the dual basis $\mathfrak{A}^* = (\omega^0, ..., \omega^3)$

$$(g(e_i, e_j))_{i,j} = \mathrm{diag}(1, -\lambda, -\lambda, -\lambda),$$
$$(h(\omega^i, \omega^j))_{i,j} = \mathrm{diag}(-\lambda, 1, 1, 1).$$

Proof. Let first $\lambda \neq 0$. Then we know from theorem (1.2.1) and from (0.3.9) that there is a basis $\mathfrak{A} = (e_0, e_1, e_2, e_3)$ of V with

$$M(g; \mathfrak{A}) = \mathrm{diag}(1, -\lambda, -\lambda, -\lambda).$$

(Actually we first get a basis $\mathfrak{A}$ with $M(g; \mathfrak{A}) = \mathrm{diag}(1, -1, -1, -1)$ for $\lambda > 0$ and $M(g; \mathfrak{A}) = \mathrm{diag}(1, 1, 1, 1)$ for $\lambda < 0$, but after scaling we can get a basis as mentioned above.) Of course, we then have:

$$(\langle \varphi(e_i), e_j \rangle) = (g(e_i, e_j)) = \mathrm{diag}(1, -\lambda, -\lambda, -\lambda).$$

We now define

$$\omega^0 := \varphi(e_0), \quad \omega^k := -\frac{1}{\lambda} \varphi(e_k) \quad (1 \leq k \leq 3)$$

and then

$$\langle \omega^i, e_j \rangle = \delta^i_j$$

holds, which means that $\mathfrak{B} := (\omega^0, \omega^1, \omega^2 \omega^3)$ is dual to $\mathfrak{A}$. Furthermore, we know that for all $v, w \in V$

$$h(\varphi(v), \varphi(w)) = -\lambda g(v, w)$$

holds and so we finally get

$$h(\omega^0, \omega^0) = -\lambda g(e_0, e_0) = -\lambda,$$
$$h(\omega^0, \omega^k) = -\frac{1}{\lambda} \cdot \lambda g(e_0, e_k) = 0,$$
$$h(\omega^k, \omega^l) = \frac{1}{\lambda^2} h(\varphi(e_k), \varphi(e_l)) = \frac{1}{\lambda^2}(-\lambda) g(e_k, e_l) = \frac{1}{\lambda^2}(-\lambda)(-\lambda) \delta_{kl} = \delta_{kl},$$

for $1 \leq k, l \leq 3$. This shows the claim for $\lambda \neq 0$.

Now let $\lambda = 0$. It is clear that we can not copy the proof for $\lambda \neq 0$ as g does not determine h in case of $\lambda = 0$. As the degeneration space of h is not as big as the one of g we now start with h:

Due to theorem (1.2.1) and (0.3.9) we now can choose a basis $\mathfrak{B} := (\omega^0, \omega^1, \omega^2, \omega^3)$ of V^* so that

$$M(h; \mathfrak{B}) = \operatorname{diag}(0, 1, 1, 1). \tag{1.2}$$

For $\mu \neq 0$, $\mathfrak{B}_\mu := (\mu \omega^0, \omega^1, \omega^2, \omega^3)$ is also a basis with (1.2) as h is bilinear.

Now let $\mathfrak{A} = (e_0, e_1, e_2, e_3)$ be dual to $\mathfrak{B}$. Then $\mathfrak{A} := (\frac{1}{\mu} e_0, e_1, e_2, e_3)$ is dual to $\mathfrak{B}_\mu$. Since

$$\langle \omega^\alpha, \psi(\omega^i) \rangle = h(\omega^\alpha, \omega^i) = \delta^{\alpha i} \quad (0 \leq \alpha \leq 3,\ 1 \leq i \leq 3),$$

we get for $1 \leq i \leq 3$:

$$\psi(\omega^i) = e_i.$$

Then it follows

$$g(e_\alpha, e_i) = g(\psi(\omega^i), e_\alpha) = \underbrace{\langle \varphi \circ \psi(\omega^i), e_\alpha \rangle}_{=0} = 0, \quad (0 \leq \alpha \leq 3,\ 1 \leq i \leq 3).$$

Now let finally $c := g(e_0, e_0)$. Then $c > 0$ as $\operatorname{rk}(g) = 1$ and $\operatorname{ind}(g) = 0$ and

$$g\left(\frac{1}{\mu} e_0, \frac{1}{\mu} e_0\right) = \frac{1}{\mu^2} c = 1$$

for $\mu := \sqrt{c}$. So $\mathfrak{A}_\mu$ is the required basis and this shows the claim. $\square$

Now we will see that the claims of the theorem above are equivalent:

Theorem 1.2.5 *Let V be a real vector space of dimension four, $g : V \times V \to \mathbb{R}$ and $h : V^* \times V^* \to \mathbb{R}$ symmetric bilinear forms. Furthermore, let $\lambda \in \mathbb{R}$. Then it is equivalent:*

(i) g and h fulfil the following conditions:

1. *There is a $v \in V$ with $g(v, v) > 0$ and $h|_{\operatorname{Ann}(v) \times \operatorname{Ann}(v)} > 0$.*

2. For
$$\varphi: V \to V^*, \ \langle \varphi(v), w \rangle = g(v, w)$$
$$\psi: V^* \to V, \ \langle \alpha, \psi(\beta) \rangle = h(\alpha, \beta),$$

where $v, w \in V$ and $\alpha, \beta \in V^*$,
$$\varphi \circ \psi = -\lambda \operatorname{id}_{V^*}$$
holds.

(ii) There is a basis $\mathfrak{A} = (e_0, ..., e_3)$ of V so that we have for the dual basis $\mathfrak{A}^* = (\omega^0, ..., \omega^3)$
$$(g(e_i, e_j))_{i,j} = \operatorname{diag}(1, -\lambda, -\lambda, -\lambda),$$
$$(h(\omega^i, \omega^j))_{i,j} = \operatorname{diag}(-\lambda, 1, 1, 1).$$

Proof.

$(i) \Rightarrow (ii)$ This directly follows from the normal form theorem (1.2.4).

$(ii) \Rightarrow (i)$ We now show that g and h satisfy the conditions demanded.

1) The first condition is fulfilled immediately. By defining $v := e_0$ and by using the assumption we have:
$$g(v, v) = 1 > 0.$$
Then the annihilator of v is:
$$\operatorname{Ann}(v) = \operatorname{Ann}(e_0) = \operatorname{span}(\omega^1, \omega^2, \omega^3)$$
and h is obviously positive definite on $\operatorname{span}(\omega^1, \omega^2, \omega^3)$.

2) If we want to see the validity of the second condition we have to present the maps
$$\varphi: V \to V^*, \ \langle \varphi(v), w \rangle = g(v, w)$$
$$\psi: V^* \to V, \ \langle \alpha, \psi(\beta) \rangle = h(\alpha, \beta)$$
with regard to the bases $\mathfrak{A}$ and $\mathfrak{A}^*$ and we get the matrices
$$A = \operatorname{diag}(1, -\lambda, -\lambda, -\lambda) \quad \text{and} \quad B = \operatorname{diag}(-\lambda, 1, 1, 1).$$
From $A \cdot B = -\lambda E_4$ follows that $\varphi \circ \psi = -\lambda \operatorname{id}_{V^*}$. This shows the second condition and the theorem.

$\square$

Remark 1.2.6 As mentioned above (see (1.1.3 h)), we now are able to show that our axiom one is equivalent to Ehlers' axioms:

1. At every point $p \in M$ there is at least one timelike tangent vector $\xi \in TM_p$.

2. For every $\xi \in TM_p$ with $g_p(\xi, \xi) > 0$,
$$h_p|_{\operatorname{Ann}(\xi) \times \operatorname{Ann}(\xi)} > 0.$$

As we showed in theorem (1.2.5) that our axioms in (i) are equivalent to (ii), we just have to show that Ehlers' second condition follows from (ii).
In order to demonstrate this we choose an arbitrary $v \in V$ with $g(v,v) > 0$ and we consider
$$\operatorname{Ann}(v) = \{\alpha \in V^* | \alpha(v) = 0\}.$$

The case of $\lambda < 0$ is not interesting as h then is positive definite on the whole of V^* and, of course, also on $\operatorname{Ann}(v)$. So we look at the case of $\lambda \geq 0$. Let $\mathfrak{A} = (e_0, ..., e_3)$ be a basis of V and $\mathfrak{A}^* = (\omega^0, ..., \omega^3)$ the dual basis so that
$$(g(e_i, e_j))_{i,j} = \operatorname{diag}(1, -\lambda, -\lambda, -\lambda),$$
$$(h(\omega^i, \omega^j))_{i,j} = \operatorname{diag}(-\lambda, 1, 1, 1).$$

Then there is a presentation of v with respect to this basis: $v = \sum_{i=0}^{3} v_i e_i$, $v_i \in \mathbb{R}$, $i = 0, ..., 3$, and we have:
$$0 < g(v,v) = (v_0)^2 - \lambda \sum_{i=1}^{3} (v_i)^2.$$

It immediately follows that $v_0 \neq 0$. If you now consider
$$0 < g(v,v) = (v_0)^2 \left(1 - \lambda \frac{\sum_{i=1}^{3}(v_i)^2}{(v_0)^2}\right),$$
you can see that $\lambda \frac{|\overline{v}|^2}{v_0^2} < 1$ for $\overline{v} := (v_1, v_2, v_3)$.
Let $\beta \in \operatorname{Ann}(v)$ be arbitrary with the presentation $\beta = \sum_{i=0}^{3} \beta_i \omega^i$, $\beta_i \in \mathbb{R}$, $i = 0, ..., 3$. Then:
$$0 = \beta(v) = \sum_{i=0}^{3} \beta_i v_i = v_0 \beta_0 + \sum_{i=1}^{3} \beta_i v_i \Leftrightarrow \beta_0 = -\frac{\sum_{i=1}^{3} \beta_i v_i}{v_0}.$$

We now define $\overline{\beta} := (\beta_1, \beta_2, \beta_3)$ and let $(\_,\_)$ be the standard scalar product on $\mathbb{R}^3$, then $(\overline{\beta}, \overline{v}) = \sum_{i=1}^{3} \beta_i v_i$, then the equation above results in
$$\beta_0 = -\frac{(\overline{\beta}, \overline{v})}{v_0}.$$

Then the following holds:

$$h(\beta,\beta) = -\lambda(\beta_0)^2 + \sum_{i=1}^{3}(\beta_i)^2 = -\lambda(\beta_0)^2 + |\overline{\beta}|^2 = -\lambda\frac{(\overline{\beta},\overline{v})^2}{(v_0)^2} + |\overline{\beta}|^2$$

$$\geq -\lambda\frac{|\overline{\beta}|^2|\overline{v}|^2}{(v_0)^2} + |\overline{\beta}|^2 = |\overline{\beta}|^2\underbrace{\left(1 - \lambda\frac{|\overline{v}|^2}{(v_0)^2}\right)}_{>0}.$$

(Here we use the Cauchy-Schwarz inequality.) Due to the fact that $\left(1 - \lambda\frac{|\overline{v}|^2}{(v_0)^2}\right) > 0$, it follows that

$$h(\beta,\beta) \geq 0 \text{ and } h(\beta,\beta) = 0 \Leftrightarrow \overline{\beta} = 0.$$

With $\beta_0 = -\frac{(\overline{\beta},\overline{v})}{v_0}$ we then have

$$h(\beta,\beta) \geq 0 \text{ and } h(\beta,\beta) = 0 \Leftrightarrow \beta = 0.$$

But this means that h is positive definite on $\operatorname{Ann}(v)$.

1.3 Transfer to the frame theory

Now we want to transfer the results of the last paragraph to our case. We just have to cite the first theorem as its statement is just pointwise and the proof is identical with the one above.

Theorem 1.3.1 *Let M be a four-dimensional connected manifold, g a symmetric tensor field on the tangent bundle of M, $g \in \Gamma(M; T^{(0,2)}M)$, h a symmetric tensor field on the cotangent bundle of M, $h \in \Gamma(M; T^{(2,0)}M)$ and $\lambda \in \mathbb{R}$. These objects are required to satisfy the first two axioms of the frame theory. Then the following holds (everywhere!):*

(i) *For $\lambda > 0$, g and h have rank 4, g has index 3 and h has index 1.*

(ii) *For $\lambda < 0$, g and h have rank 4 and both have index 0.*

(iii) *For $\lambda = 0$, g has rank 1 and h has rank 3. Both have index 0.*

Theorem 1.3.2 *Let M be a four-dimensional connected manifold, g a symmetric tensor field on the tangent bundle of M, $g \in \Gamma(M; T^{(0,2)}M)$, h a symmetric tensor field on the cotangent bundle of M, $h \in \Gamma(M; T^{(2,0)}M)$ and $\lambda \in \mathbb{R}$. Then it is equivalent:*

(i) *g and h satisfy the first two axioms of the frame theory.*

(ii) *For every $p \in M$ there is an open neighbourhood $U \subseteq M$ and vector fields $X_i \in \mathfrak{X}(U)$, $i = 0, ..., 3$, forming a basis at every point $q \in U$ so that for the dual differential forms $\omega^j \in \mathcal{E}^{(1)}(U)$, $j = 0, ..., 3$, the following holds:*

$$(g(X_i, X_j))_{i,j} = \operatorname{diag}(1, -\lambda, -\lambda, -\lambda), \quad i,j = 0,...,3;$$
$$(h(\omega^i, \omega^j))_{i,j} = \operatorname{diag}(-\lambda, 1, 1, 1), \quad i,j = 0,...,3.$$

Proof.

$(i) \Rightarrow (ii)$ We already now that the theorem holds for one $p \in M$. We therefore have to show that is also holds locally. We first choose a chart

$$x : U \to V \subseteq \mathbb{R}^4$$

so that the coordinate vector fields

$$\left(\frac{\partial}{\partial x^0}\Big|_p, \frac{\partial}{\partial x^1}\Big|_p, \frac{\partial}{\partial x^2}\Big|_p, \frac{\partial}{\partial x^3}\Big|_p \right)$$

and the dual forms

$$(dx^0|_p, dx^1|_p, dx^2|_p, dx^3|_p)$$

are "normal bases" in p.

Since $\left(\frac{\partial}{\partial x^0}|_q, \frac{\partial}{\partial x^1}|_q, \frac{\partial}{\partial x^2}|_q, \frac{\partial}{\partial x^3}|_q \right)$ and the dual forms $(dx^0|_q, dx^1|_q, dx^2|_q, dx^3|_q)$ are bases of TM_q and TM_q^* in every $q \in U$, we now use an orthonormalization process which depends differentiable on $q \in U'$, for all $q \in U' \subseteq U$, $p \in U'$ and U' small enough (similar to the Gram-Schmidt-process).

So let us start with the frame $\left(\frac{\partial}{\partial x^0}, \frac{\partial}{\partial x^1}, \frac{\partial}{\partial x^2}, \frac{\partial}{\partial x^3} \right)$. Since

$$g_p \left(\frac{\partial}{\partial x^0}\Big|_p, \frac{\partial}{\partial x^0}\Big|_p \right) = 1,$$

we can find an open neighbourhood U' of p, $U' \subseteq U$ so that

$$g \left(\frac{\partial}{\partial x^0}\Big|_{U'}, \frac{\partial}{\partial x^0}\Big|_{U'} \right) \neq 0.$$

Then we can define

$$X_0' := \frac{1}{\sqrt{g \left(\frac{\partial}{\partial x^0}|_{U'}, \frac{\partial}{\partial x^0}|_{U'} \right)}} \cdot \frac{\partial}{\partial x^0}\Big|_{U'}.$$

Thus,

$$g(X_0', X_0') = 1$$

on U'. Now we define

$$X_i' := \frac{\partial}{\partial x^i}\Big|_{U'} - g\left(X_0', \frac{\partial}{\partial x^i}\Big|_{U'} \right)$$

for $1 \leq i \leq 3$ and it follows that

$$g(X_i', X_0') = 0.$$

We therefore consider the frame

$$\mathfrak{A}' = (X_0', X_1', X_2', X_3')$$

and its dual frame
$$'\mathfrak{A}^* = ('\omega^0, '\omega^1, '\omega^2, '\omega^3).$$

Then $\varphi(X_0') = {'\omega^0}$ holds, since:
$$\langle \varphi(X_0'), X_i' \rangle = g(X_0', X_i') = \delta_{0i},$$

for $0 \leq i \leq 3$; but we also have
$$\langle '\omega^0, X_i' \rangle = \delta_i^0,$$

since $'\mathfrak{A}^*$ is dual to $\mathfrak{A}'$. Thus, it follows that
$$h('\omega^0, '\omega^i) = \langle '\omega^i, \psi('\omega^0) \rangle = \langle '\omega^i, \psi(\varphi(X_0')) \rangle$$
$$= -\lambda \langle '\omega^i, X_0' \rangle = -\lambda \delta_0^i,$$

for $0 \leq i \leq 3$. Now we consider
$$W := \{\omega \in \mathcal{E}^{(1)}(U') : \langle \omega, X_0' \rangle = 0\}.$$

Thus, for all $p \in U'$
$$W_p = \mathrm{span}('\omega^1|_p, '\omega^2|_p, '\omega^3|_p) = \mathrm{Ann}(X_0'|_p)$$

holds and $('\omega^1, '\omega^2, '\omega^3)$ is a basis of TM_p^* in every $p \in U'$. Due to the first condition, which is equivalent to Ehlers' axioms as we have seen in (1.2.6, we now know that h_p is positive definite on the annihilator of $X_0'|_p$ and therefore it follows that
$$h('\omega^i, '\omega^i) > 0,$$

for $1 \leq i \leq 3$. Therefore, we now are able to orthonormalize $('\omega^1, '\omega^2, '\omega^3)$ using the standard orthonormalization process of Schmidt. We then get $\omega^i \in \mathcal{E}^{(1)}(U')$ with
$$h(\omega^i, \omega^j) = \delta^{ij}$$

for $1 \leq i, j \leq 3$. Thus, we define
$$\mathfrak{A}^* := (\omega^0, \omega^1, \omega^2, \omega^3)$$

with $\omega^0 := {'\omega^0}$ and we already have
$$(h(\omega^i, \omega^j))_{i,j} = \mathrm{diag}(-\lambda, 1, 1, 1).$$

1.3. Transfer to the frame theory

Now let $\mathfrak{A} = (X_0, X_1, X_2, X_3)$ be dual to $\mathfrak{A}^*$, thus

$$\langle \omega^j, X_i \rangle = \delta_i^j,$$

for $0 \leq i, j \leq 3$. But this means that $X_0 = X_0'$, since

$$\langle \omega^j, X_0' \rangle = \delta_0^j = \langle \omega^j, X_0 \rangle,$$

for $0 \leq j \leq 3$. Furthermore, we have

$$g(X_0, X_i) = \langle \varphi(X_0), X_i \rangle = \langle \omega^0, X_i \rangle = \delta_i^0.$$

Next we show that $\psi(\omega^i) = X_i$ for $1 \leq i \leq 3$. This follows from the fact that

$$\langle \omega^j, \psi(\omega^i) \rangle = h(\omega^i, \omega^j) = \delta^{ij} = \delta_i^j = \langle \omega^j, X_i \rangle,$$

for $0 \leq j \leq 3$. Finally,

$$\begin{aligned} g(X_i, X_j) &= g(\psi(\omega^i), X_j) = \langle \varphi(\psi(\omega^i)), X_j \rangle \\ &= -\lambda \langle \omega^i, X_j \rangle = -\lambda \delta_j^i, \end{aligned}$$

for $1 \leq i, j \leq 3$, and therefore

$$(g(X_i, X_j))_{i,j} = \mathrm{diag}(1, -\lambda, -\lambda, -\lambda),$$

which shows the claim.

$(ii) \Rightarrow (i)$ We now show that the first two axioms of the frame theory hold. Let (X_0, X_1, X_2, X_3) and $(\omega^0, \omega^1, \omega^2, \omega^3)$ so that

$$\begin{aligned} (g(X_i, X_j))_{i,j} &= \mathrm{diag}(1, -\lambda, -\lambda, -\lambda), \quad i, j = 0, ..., 3; \\ (h(\omega^i, \omega^j))_{i,j} &= \mathrm{diag}(-\lambda, 1, 1, 1), \quad i, j = 0, ..., 3. \end{aligned}$$

1) Then, the first axiom is satisfied immediately, since we take $X_0|_p$ for all $p \in U'$ and get

$$g_p(X_0|_p, X_0|_p) = 1.$$

Furthermore,

$$\mathrm{Ann}(X_0|_p) = \mathrm{span}(\omega_p^1, \omega_p^2, \omega_p^3),$$

where h_p is obviously positive definite.

2) Since this is also a pointwise discussion the second axiom follows directly from the proof of (1.2.5) and this shows the claim of the theorem.

$\square$

Remark 1.3.3 (i) The arguments used in the proof above show why it is usually not possible to prove the theorem globally. A global theorem would require a global chart x and thus a global frame on M. If this was the case one could start this orthonormalization process and get a global assertion.

(ii) If you want to use the frame theory to make a connection between General Relativity and Newton's theory of gravitation you first have to find out how you can detect the two theories within the frame theory. In case of $\lambda > 0$, models of General Relativity can appear. Then we have a four-dimensional manifold M, $(-\lambda g)$ is of full rank and has index 1, so it is a Lorentzian metric and, in addition to that, Einstein's field equations hold according to the sixth axiom of the frame theory. Up to this moment it is not clear at all how you can discover models of Newton's theory within the frame theory. As we will soon see, these models can appear in case of $\lambda = 0$. This is the topic of the following section.

1.4 The case $\lambda = 0$

In this section we now discuss the case $\lambda = 0$. We will see that Newtonian models can appear in this case. Of course, this is not a new fact, see [31] for instance. But as already mentioned before, we are interested in the global structure of the manifold, not in the local one, as in the approaches by Lottermoser, for instance. This global point of view would be an improvement concerning the knowledge about the frame theory up to this moment.

As we have already noted before in case of $\lambda > 0$, models of General Relativity can appear. The connection ∇ is totally and unambiguously defined. Some of the axioms also follow directly from others (see (1.1.3)). If we want to detect Newton's theory in case of $\lambda = 0$, we first have to show that the manifold M equals $M = \mathbb{R} \times \mathbb{R}^3$ in this case.

We therefore consider M, g, h and ∇ as well as the axioms belonging to them in case of $\lambda = 0$. From theorem (1.3.1) follows that $\operatorname{rk}(g) = 1$, $\operatorname{ind}(g) = 0$, $\operatorname{rk}(h) = 3$ and $\operatorname{ind}(h) = 0$. Therefore, we now define

$$H_p := \operatorname{Deg}(g_p) = \{\xi \in TM_p : g_p(\xi, \eta) = 0, \ \forall \eta \in TM_p\} \subseteq TM_p.$$

Because of $\operatorname{rk}(g) = 1$ we have $\dim(H_p) = \dim(\operatorname{Deg}(g_p)) = 3$. Furthermore, we define $H := (H_p)_{p \in M}$. Then H is a subset of TM, $H \subseteq TM$ and as H_p depends smoothly on p, H is a smooth subbundle of TM. Therefore, $H = (H_p)_{p \in M}$ is a distribution on M (see also (1.4.1)).

1.4.1 The theorem of Frobenius

As we need the theorem of Frobenius in the following discussion we now cite this theorem and the definitions belonging to this context. If you are interested in further reading see [44], chapter 1, for instance.

Definition 1.4.1 Let M^n be a smooth manifold and let $0 \leq k \leq n$.

a) A *distribution of rank k* on M is a smooth subbundle $E \subseteq TM$ of rank k.

b) A distribution $E \subseteq TM$ is called *integrable*, if for every $p \in M$ there is a chart $x : U \to V \subseteq \mathbb{R}^n$ (with $p \in U$, $0 \in V$ and $x(p) = 0$) so that

$$N_c = \{q \in U : x^{k+1}(q) = c^{k+1}, ..., x^n(q) = c^n\}$$

is an *integral manifold* for E, for all $c \in V \subseteq \mathbb{R}^{n-k}$, where $V \subseteq \mathbb{R}^{n-k}$ is a neighbourhood of $0 \in \mathbb{R}^{n-k}$. This means that $(TN_c)_q = E_q$, for all $q \in N_c$.

Definition 1.4.2 We call a distribution $E \subseteq TM$ *involutive*, if

$$[X, Y] \in \Gamma(E)$$

holds for all $X, Y \in \Gamma(E)$.

Theorem of Frobenius 1.4.3 *A distribution $E \subseteq TM$ is integrable if and only if it is involutive.*

1.4.2 The topology of (M, H)

In order to achieve $M = \mathbb{R} \times \mathbb{R}^3$ we first have to show that the distribution is integrable and then that the foliation even gives a fibration. If we then show in addition that the Frobenius leaves equal $\mathbb{R}^3$ and are not twisted we meet our aim. So let us start dealing with integrability.

Remark 1.4.4 (i) Locally the foliation is always a product, of course, globally it can be very complicated. As we will see in the following discussion we can only achieve global status if the manifold meets further conditions, if it is simply connected, for instance.

(ii) In the following we will show that some results which we know from semi-Riemannian geometry, for instance, also hold in case of $\lambda = 0$. So we develop a special geometry for degenerate bilinear forms.

(iii) As the connection ∇ is also symmetric and metric with regard to g in case of $\lambda = 0$ the Koszul formula also holds:

$2g(\nabla_X Y, Z)$
$= X(g(Y, Z)) + Y(g(Z, X)) - Z(g(X, Y)) - g(X, [Y, Z]) + g(Y, [Z, X]) + g(Z, [X, Y])$,

for X, Y and $Z \in \mathfrak{X}(M)$. This can easily be seen because one just has to apply symmetry and metric property repeatedly:

$$\begin{aligned}
g(\nabla_X Y, Z) &= X(g(Y,Z)) - g(Y, \nabla_X Z) = X(g(Y,Z)) - (g(Y, \nabla_Z X) + g(Y, [X,Z]))\\
&= X(g(Y,Z)) - (Z(g(Y,X)) - g(\nabla_Z Y, X)) - g(Y, [X,Z])\\
&= X(g(Y,Z)) - Z(g(Y,X)) + g(\nabla_Y Z + [Z,Y], X)) - g(Y, [X,Z])\\
&= X(g(Y,Z)) - Z(g(Y,X)) + Y(g(Z,X)) - g(Z, \nabla_Y X) + g(X, [Z,Y])\\
&\quad + g(Y, [Z,X])\\
&= X(g(Y,Z)) + Y(g(Z,X)) - Z(g(X,Y)) - g(Z, \nabla_X Y + [Y,X]) - g(X, [Y,Z])\\
&\quad + g(Y, [Z,X])\\
&= X(g(Y,Z)) + Y(g(Z,X)) - Z(g(X,Y)) - g(X, [Y,Z]) + g(Y, [Z,X])\\
&\quad + g(Z, [X,Y]) - g(\nabla_X Y, Z),
\end{aligned}$$

for X, Y and $Z \in \mathfrak{X}(M)$. But you have to bear in mind that the connection ∇ is not determined unambiguously because of the degeneration of g.

(iv) We want to mention again that in the tangent and cotangent spaces in case of $\lambda = 0$ we have

$$H_p := \ker(\varphi_p) = \mathrm{Deg}(g_p) = \mathrm{im}(\psi_p) \subseteq TM_p,$$
$$L_p := \ker(\psi_p) = \mathrm{Deg}(h_p) = \mathrm{im}(\varphi_p) \subseteq TM_p^*.$$

Furthermore, $\dim(H_p) = 3$ and $\dim(L_p) = 1$.

Theorem 1.4.5 *Let $H = (H_p)$ be the distribution defined above with $H_p = \mathrm{Deg}(g_p)$. Then the distribution $H = (H_p)$ is integrable.*

Proof. First, we have to remember that the integrability of the distribution $H \subseteq TM$ is a local condition. Therefore, let $p \in M$, $x : U \to V \subseteq \mathbb{R}^4$ be a chart for p so that $\left(\frac{\partial}{\partial x^0}|_p, \frac{\partial}{\partial x^1}|_p, \frac{\partial}{\partial x^2}|_p, \frac{\partial}{\partial x^3}|_p\right)$ is a normal basis (see (1.3.2)). We now show that the distribution is involutive. Then the theorem of Frobenius provides that it is also integrable. In order to achieve this we use $B := \frac{\partial}{\partial x^0} \in \mathfrak{X}(U')$, with $p \in U' \subseteq U$ and U' small enough. Then

$$g(B, B) = 1$$

on U'. We then define

$$\omega := \varphi(B)(\_) = g(B, \_).$$

Then $\omega_p \neq 0$ because of the fact that $g_p(B_p, B_p) = 1$ for all $p \in U'$. Furthermore, we have

$$\ker(\omega_p) = \{\xi \in TM_p : 0 = \langle \varphi_p(B_p), \xi \rangle = g_p(B_p, \xi)\}.$$

Then $H_p \subseteq \ker(\omega_p)$ and because of $\omega_p \neq 0$ for all $p \in U'$, follows that $\ker(\omega_p) = H_p$ due to dimension.

In order to show that H is involutive, let $X, Y \in \Gamma(H)$ be arbitrary and then we have to prove that $[X, Y] \in \Gamma(H)$. We just have to show that $\omega([X, Y]) = 0$ because this means that $[X, Y]_p \in \ker(\omega_p)$ and we get $[X, Y] \in \Gamma(H)$.

By definition of ω we have $\omega([X, Y]) = g(B, [X, Y])$. As ∇ is symmetric and metric with regard to g it follows from the Koszul formula (see (1.4.4)) that

$$\omega([X, Y]) = g(B, [X, Y]) = -2\underbrace{g(\nabla_X B, Y)}_{=0} + X\underbrace{g(B, Y)}_{=0} + B\underbrace{g(X, Y)}_{=0}$$
$$- Y\underbrace{g(X, B)}_{=0} - \underbrace{g(X, [B, Y])}_{=0} + \underbrace{g(Y, [X, B])}_{=0} = 0,$$

as $X, Y \in \Gamma(H)$ (remember: $\text{Deg}(g_p) = H_p$). So H is involutive and therefore also integrable by the theorem of Frobenius. □

Remark 1.4.6 If we have a look at the definition of ω in the proof of (1.4.5) it seems to depend on the choice of B. But, while the choice of B is arbitrary, ω is not arbitrary at all: $\omega = \varphi(B)$, so $\omega \in L$ and $h_p(\omega_p, \omega_p) = 1$. As $\dim(L_p) = 1$, ω is fixed except for its sign.

In the following we want to discuss the global structure of the foliation. We therefore suppose that there is a global timelike normalized vector field $B \in \mathfrak{X}(M)$. Up to this moment the existence of such a vector field is not clear at all and has to be studied in the following. So let us suppose that $B \in \mathfrak{X}(M)$ is such a vector field. Then we first note that

$$0 = \nabla_\xi(g(B, B)) = (\nabla_\xi g)(B_p, B_p) + g_p(\nabla_\xi B, B_p) + g_p(B_p, \nabla_\xi B) = 2g_p(\nabla_\xi B, B_p),$$

for $\xi \in TM_p$. Therefore,

$$g_p(\nabla_\xi B, B_p) = 0 \tag{1.3}$$

for $\xi \in TM_p$ arbitrary.

Furthermore, let $X \in \Gamma(H)$. Then:

$$g_p(\nabla_{B_p} X, B_p) = B_p(\underbrace{g(X, B)}_{=0}) - \underbrace{(\nabla_{B_p} g)}_{=0}(X_p, B_p) - \underbrace{g_p(X_p, \nabla_{B_p} B)}_{=0} = 0. \tag{1.4}$$

If we now again define $\omega = \langle \varphi(B), \_ \rangle = g(B, \_)$, we can show that ω is closed:

$$d\omega = 0.$$

As $d\omega_p$ is an antisymmetric bilinear form, we just have to show that

$$d\omega_p(\xi, \eta) = 0$$

for $\xi, \eta \in H_p$ and

$$d\omega_p(\xi, B_p) = 0$$

for $\xi \in H_p$. Now let $\xi, \eta \in H_p$ be arbitrary and $X, Y \in \Gamma(H)$ continuations of ξ and η, $X_p = \xi$,

$Y_p = \eta$. Then it follows with the help of the *Formula of Cartan*:

$$d\omega_p(\xi, \eta) = \xi(\underbrace{\omega(Y)}_{=0}) - \eta(\underbrace{\omega(X)}_{=0}) - \omega_p(\underbrace{[X, Y]_p}_{\in H_p}) = 0.$$

Due to (1.3) and (1.4) we also get

$$d\omega_p(\xi, B_p) = \xi(\underbrace{\omega(B)}_{=1}) - B_p(\underbrace{\omega(X)}_{=0}) - \omega_p([X, B]_p) = \omega_p(\nabla_\xi B - \nabla_{B_p} X)$$
$$= g_p(B_p, \nabla_\xi B) - g_p(B_p, \nabla_{B_p} X) = 0.$$

Therefore,
$$d\omega = 0.$$

Notation 1.4.7 In the following discussion the foliation belonging to H is called $\mathcal{B} = (F_\alpha)_{\alpha \in R}$, where R is an index set. This means that the F_α are maximal and connected integral manifolds for H and $M = \dot{\bigcup}_{\alpha \in R} F_\alpha$.

Definition 1.4.8 Let M and N be smooth manifolds. A smooth map $\Phi : M \to N$ is called submersion, if $D\Phi_p : TM_p \to TN_{\Phi(p)}$ is surjective for all $p \in M$.

Theorem 1.4.9 *Let the manifold M be simply connected and g and ∇ be as above. Suppose there is a complete global timelike normalized vector field $B \in \mathfrak{X}(M)$ on M. Then there is a submersion $T : M \to \mathbb{R}$ so that $T^{-1}(c)$ is an integral manifold for H for every $c \in \mathbb{R}$.*

Proof. We first define the differential form $\omega \in \mathcal{E}^{(1)}(M)$ by

$$\omega := \langle \varphi(B), \_ \rangle = g(B, \_).$$

According to (1.4.6), we already know that ω then is closed, $d\omega = 0$. As the manifold M is simply connected, $\pi_1(M) = (1)$, it follows that ω is also exact (see for example [40], §5.7). Therefore, there is a $T \in \mathcal{E}(M)$ for ω with $dT = \omega$. Because of $dT_p(B_p) = \omega_p(B_p) = 1$ for all $p \in M$, T is a submersion.

Now let $(\Phi^t : M \to M)_{t \in \mathbb{R}}$ be the global flow on M belonging to B, this means that $\Phi^0 = \text{id}$ and
$$\frac{d\Phi^t}{dt} = B \circ \Phi^t,$$
for all $t \in \mathbb{R}$. For all $p \in M$ we then have

$$\frac{d}{dt}(T \circ \Phi^t(p)) = \langle dT_{\Phi^t(p)}, \frac{d\Phi^t}{dt}(p)\rangle = \langle \omega_{\Phi^t(p)}, B_{\Phi^t(p)}\rangle = 1,$$

which means that
$$T \circ \Phi^t(p) = T(\Phi^0(p)) + t = T(p) + t. \tag{1.5}$$

Therefore, it follows that
$$T(M) = \mathbb{R}.$$

As T is a submersion, the surfaces
$$N_c := T^{-1}(c) \subseteq M$$

are hypersurfaces (closed submanifolds of codimension 1). Furthermore, for all $p \in N_c$
$$T_p(N_c) = \ker(dT_p) = \ker(\omega_p) = H_p$$

holds, which means that $N_c = T^{-1}(c)$ is integral manifold for H. $\square$

Theorem 1.4.10 *Let the manifold M be simply connected and g and ∇ be as above. Suppose there is a complete global timelike normalized vector field $B \in \mathfrak{X}(M)$ on M. Furthermore, let $T: M \to \mathbb{R}$ be the submersion of theorem (1.4.9). Then there is a diffeomorphism $\Psi : \mathbb{R} \times F \to M$ with $T \circ \Psi = pr_1$, where $F := T^{-1}(0) = N_0$.*

Proof. Let $F := T^{-1}(0) = N_0$. We define
$$\Psi : \mathbb{R} \times F \to M, \quad \Psi(t,p) := \Phi^t(p),$$

where $(\Phi^t(p))_{t \in \mathbb{R}}$ is the global flow belonging to B. Then Ψ is smooth and the same holds for
$$\Omega : M \to \mathbb{R} \times F, \quad \Omega(p) := (T(p), \Phi^{-T(p)}(p)).$$

According to (1.5), we know that
$$T(\Phi^t(p)) = T(p) + t,$$

for all $t \in \mathbb{R}$. Therefore, the following holds:
$$\Omega \circ \Psi(t,p) = \Omega(\Phi^t(p)) = (T(\Phi^t(p)), \Phi^{-T(\Phi^t(p))}(\Phi^t(p))) = (T(p) + t, \Phi^{-T(p)-t}(\Phi^t(p))).$$

Now we use the both fact that $T(p) = 0$ and the flow property $\Phi^t \circ \Phi^s = \Phi^{t+s}$, for all $s,t \in \mathbb{R}$. Then it holds:
$$\Omega \circ \Psi(t,p) = (T(p) + t, \Phi^{-T(p)-t}(\Phi^t(p))) = (t, \Phi^{-t} \circ \Phi^t(p)) = (t, \phi^0(p)) = (t,p).$$

We then also have:
$$\Psi \circ \Omega(p) = \Psi(T(p), \Phi^{-T(p)}(p)) = \Phi^{T(p)}(\Phi^{-T(p)}(p)) = \Phi^{T(p)} \circ \Phi^{-T(p)}(p) = \Phi^0(p) = p,$$

which shows that $\Omega \circ \Psi = \mathrm{id}$ and $\Psi \circ \Omega = \mathrm{id}$ and therefore: Ψ is a diffeomorphism. Finally we have
$$T \circ \Psi(t,p) = T(\Phi^t(p)) = T(p) + t = t = \mathrm{pr}_1(t,p),$$
for all $(t,p) \in \mathbb{R} \times F$, thus $T \circ \Psi = \mathrm{pr}_1$, which shows the claim. □

Corollary 1.4.11 *The surfaces $N_c := T^{-1}(c)$ are connected and simply connected.*

Proof. From theorem (1.4.10) we already know that M is diffeomorphic to $\mathbb{R} \times F$, $M \cong \mathbb{R} \times F$. As the manifold M is connected and simply connected (and $\mathbb{R}$ also, of course), $F = T^{-1}(0) = N_0$ has to be connected and simply connected. Due to $M \cong \mathbb{R} \times F$, the surfaces N_c are diffeomorphic and therefore N_c is connected and simply connected for all $c \in \mathbb{R}$. □

Theorem 1.4.12 *Let the manifold M be simply connected and g and ∇ be as above. Suppose there is a complete global timelike normalized vector field $B \in \mathfrak{X}(M)$ on M. Let $T : M \to \mathbb{R}$ be the submersion of (1.4.9). Then $\{T^{-1}(c)\}_{c \in \mathbb{R}}$ is the Frobenius foliation of H. This means that for every $c \in \mathbb{R}$ there is precisely one $\alpha \in R$ with $T^{-1}(c) = F_\alpha$ and for every $\alpha \in R$ there is precisely one $c \in \mathbb{R}$ with $F_\alpha = T^{-1}(c)$.*

Proof. We already know from (1.4.9) that the N_c are integral manifolds for H. Furthermore, we know that the hypersurfaces N_c are connected (see (1.4.11)). As N_c is integral and connected there is precisely one $\alpha \in R$ so that $N_c \subseteq F_\alpha$. So we just have to show that N_c is maximal and it follows that $N_c = F_\alpha$.

Therefore, let $p \in N_c \subseteq F_\alpha$ and $q \in F_\alpha$ be arbitrary. Now we want to show that $q \in N_c$. As F_α is connected (and therefore path connected), there is a curve $\gamma : [0,1] \to F_\alpha$ which connects p with q, so we have $\gamma(0) = p$ and $\gamma(1) = q$. For γ the following holds:
$$\dot\gamma(t) \in (TF_\alpha)_{\gamma(t)} = H_{\gamma(t)} = TN_{T\circ\gamma(t)} = \ker(DT_{\gamma(t)}).$$
So we get
$$0 = DT(\dot\gamma(t)) = \frac{d}{dt}(T \circ \gamma)(t),$$
and therefore
$$T \circ \gamma(t) = \mathrm{const.} = T \circ \gamma(0) = T(p) = c,$$
as well as
$$T(q) = T(\gamma(1)) = c.$$
This shows that
$$q \in T^{-1}(c) = N_c.$$
As every arbitrary point $q \in F_\alpha$ is also contained in N_c, N_c is maximal and coincides with the Frobenius leave, $N_c = F_\alpha$. So the map
$$\mathbb{R} \to R, \ c \mapsto \alpha$$

is bijective, since
$$\dot\bigcup_{c\in\mathbb{R}} N_c = M = \dot\bigcup_{\alpha\in\mathbb{R}} F_\alpha.$$
This shows the claim. □

Remark 1.4.13 a) Note that the fibration $T : M \to \mathbb{R}$ basically does not depend on the choice of B (see also (1.4.6)). If $\tilde{B}$ is another global timelike vector field, then there is an $X \in \Gamma(H)$ so that
$$\tilde{B} = \pm B + X,$$
since the set $\{\xi \in TM_p : g_p(\xi,\xi) = 1\}$ consists of the two hypersurfaces $B_p + H_p$ and $-B_p + H_p$. For $\tilde{\omega}$ we then either have
$$\tilde{\omega} := \phi(\tilde{B}) = \omega$$
or
$$\tilde{\omega} := \phi(\tilde{B}) = -\omega$$
and therefore for a $\tilde{T}$ with $D\tilde{T} = \tilde{\omega}$
$$\tilde{T} = \pm T + \tilde{c}$$
for a $\tilde{c} \in \mathbb{R}$. T is considered to be the *absolute time* of (M, g, ∇).

b) The structure of the foliation beloning to $H = \text{Deg}(g)$ of (M, g, ∇) is therefore completely clear if M is simply connected and if there is a complete vector field $B \in \mathfrak{X}(M)$ on (M, g). But we want to point out that you can always find a product structure locally. So if you do not insist on the fact that the manifold has to equal $\mathbb{R} \times \mathbb{R}^3$ but $I \times U$, where $I \subseteq \mathbb{R}$ is an open interval and $U \subseteq \mathbb{R}^3$ a connected subset, you do not have to require anything, since the theorems always hold locally.

c) In the following we have to discuss the geometrical and topological properties of the three manifold F. We already know that it is simply connected and connected.

d) Furthermore, we have to find out if there is a global timelike vector field B on (M, g, ∇), which means if (M, g) is time oriented. And we also have to discuss the question if there is a complete vector field.

Remark 1.4.14 The existence of such a complete vector field $B \in \mathfrak{X}(M)$ is a necessary condition for a global product structure. We construct a counterexample:
Let $M_0 := \mathbb{R} \times \mathbb{R}^3$ and g the "standard structure" given by $\text{diag}(1,0,0,0)$ with respect to cartesian coordinates (t, x^1, x^2, x^3). Let furthermore $\nabla = D$ be the "standard connection". Let
$$E := \{(t,x) \in M_0 : t \geq 0, x^1 = 0\}$$

and we define $M := M_0 \backslash E$.

Then $B := \frac{\partial}{\partial t}$ is a timelike unit global vector field on (M, g) but it is not complete. For the submersion $T : M \to \mathbb{R}$ belonging to B, $T = \text{pr}_1$ holds (except for an additive constant). But $T^{-1}(0)$ is not connected!

Furthermore, there is no other submersion whose fibers are the Frobenius leaves. If we define $R := M/\sim$, where $p \sim q$ if p and q belong to the same leave and $f : M \to N$ is such a submersion, R has to be at least Hausdorff. But one can show that the two leaves in $T^{-1}(0)$ are not separable in R.

1.4.3 Time-orientation of (M, H)

In this section we discuss the existence of a global timelike normalized vector field on (M, g, ∇). We therefore first prove a well known theorem (especially in Algebraic Topology, see [23], for instance). The existence of the global timelike normalized vector field in our case then follows from this theorem.

Theorem 1.4.15 *Every line bundle $L \to M$ on a simply connected manifold M is trivial.*

Proof. We here give a proof using methods of Differential Geometry.

Let M be simply connected and $L \to M$ a line bundle on M. Let h be a bundle metric on L and ∇ a metric connection with respect to h. We now show that ∇ is flat.

Because of $\dim L_p = 1$ there is a $v \in L_p \backslash \{0\}$ with $L_p = \langle v \rangle$, for all $p \in M$. As the curvature tensor $R_p(\xi, \eta) : L_p \to L_p$ is linear, $\xi, \eta \in TM_p$, we therefore just have to show that

$$R_p(\xi, \eta)v = 0,$$

if we want to show that $R_p(\xi, \eta) = 0$. But it is equivalent to show that

$$0 = h(R_p(\xi, \eta)v, v) = \text{Rm}_p(\xi, \eta, v, v).$$

But as the Riemannian curvature tensor Rm_p is anti-symmetric in the last two arguments (see (0.2.4)) this condition is fulfilled. But this exactly means that ∇ is flat.

We have shown that there is a flat connection on L. So we have almost reached our aim to show that $L \cong M \times \mathbb{R}$. Due to dimension it is already clear that every fiber L_p is isomorph to $\mathbb{R}$, $L_p \cong \mathbb{R}$, $\forall p \in M$. Now we use the fact that the manifold M is simply connected. This means that two smooth curves $\alpha : I \to M$ and $\beta : I \to M$, ($I = [0, 1]$) with fixed ending points, for example $\alpha(0) = p = \beta(0)$, $\alpha(1) = q = \beta(1)$, are homotopic. Now we employ parallel transport along α. But as we have a flat connection parallel transport is invariant with regard to homotopy. That means that it does not depend on the choice of the curve with starting point p and ending point q (see for example [2]).

Therefore, let p and $q \in M$ be arbitrary. We consider the isomorphism

$$I_q : \mathbb{R} \to L_q, \quad v \mapsto \text{par}_\alpha(\iota(v)),$$

where $\alpha : I \to M$ is an arbitrary smooth curve with $\alpha(0) = p$ and $\alpha(1) = q$ and ι is some fixed isomorphism between $\mathbb{R}$ and F_p. This now provides the demanded isomorphism Φ between $M \times \mathbb{R}$ and L by

$$\Phi : M \times \mathbb{R} \to L, \ (q, v) \mapsto I_q(v),$$

since parallel transport $\mathrm{par}_\alpha : L_p \to L_q$ is an isomorphism (even an isometry if ∇ is metric), see for example [28], §3.2. But this means that $L \cong M \times \mathbb{R}$ and therefore, that the bundle L is trivial. $\square$

Theorem 1.4.16 *Let M be a simply connected manifold, H our distribution as well as g and ∇ as above. Then there is a global timelike normalized vector field $B \in \mathfrak{X}(M)$ on M.*

Proof. In order to show this we have to look at the bundle $L = TM/H$. Due to dimension L is a line bundle on M. According to theorem (1.4.15) this bundle is trivial as the manifold M is simply connected. But as the bundle is trivial, there is a section $\sigma : M \to L$ which is not zero at any point of M. Now let γ be a Riemannian metric on M and we consider

$$F := H^\perp \subseteq TM$$

with respect to γ. If $\pi : TM \to L = TM/H$ denotes the projection and $\iota : F \to TM$ is the inclusion, then $f := \pi \circ \iota : F \to L$ is obviously a bundle isomorphism (since $f_p : F_p \to L_p$ is an isomorphism between one dimensional vector spaces). Now we can define

$$\widetilde{B} := f^{-1} \circ \sigma : M \to F \subseteq TM.$$

So we have found a global timelike vector field $\widetilde{B}$ on M, this means that $g(\widetilde{B}, \widetilde{B}) \neq 0$ everywhere. But then

$$B := \frac{1}{\sqrt{g(\widetilde{B}, \widetilde{B})}} \widetilde{B}$$

is a global timelike normalized vector field on M. $\square$

Remark 1.4.17 We therefore know that there is a global timelike normalized vector field if we require that the manifold is simply connected. But one has to bear in mind that the completeness of such a vector field can not be concluded (see (1.4.14)). But the completeness of at least one vector field is a necessary and sufficient condition for a global product structure. We therefore have to demand that one of the global timelike normalized vector fields is complete if we want to have a global product structure.

The existence of such a global timelike vector field is a further improvement of the facts known up to this moment. Lottermoser, for instance, always requires the existence of such a vector field. Now we have shown that it always exists if the manifold M is simply connected.

1.4.4 Geometry of (M, H)

We now know the global structure of the fibration. Therefore, we have to discuss the structure of the leaves N_c for every c. But as we already know that the leaves are diffeomorphic we just discuss the case $F := N_0$.

First, we will show that F is flat. So let us consider $h_p : TM_p^* \times TM_p^* \to \mathbb{R}$ and the map $\psi_p : TM_p^* \to TM_p \cong TM_p^{**}$ (canonical) with $\langle \beta, \psi_p(\alpha) \rangle = h_p(\alpha, \beta)$. We already know that $\mathrm{ind}(h_p) = 3$ and $\mathrm{rk}(h_p) = 3$, thus $\dim(\mathrm{Deg}(h_p)) = 1$. As $\ker(\psi_p) = \mathrm{Deg}(h_p)$, ψ_p induces a homomorphism $\widetilde{\psi}_p : D_p := TM_p^*/\mathrm{Deg}(h_p) \to TM_p$. Since $\mathrm{im}(\psi_p) = H_p$, we already know that $\mathrm{im}(\widetilde{\psi}_p) \subseteq H_p$. But as $\ker(\widetilde{\psi}_p) = (0)$ it follows that $\dim(\mathrm{im}(\widetilde{\psi}_p)) = 3$ and therefore $\mathrm{im}(\widetilde{\psi}_p) = H_p$. Thus,

$$\widetilde{\psi}_p : D_p = TM_p^*/\mathrm{Deg}(h_p) \stackrel{\cong}{\to} H_p$$

is an isomorphism.

Now, $h_p : TM_p^* \times TM_p^* \to \mathbb{R}$ also induces $h'_p : D_p \times D_p \to \mathbb{R}$ and since h_p is bilinear, symmetric and positive semi-definite, h'_p is also. But h'_p is even positive definite, since from $h'_p([\alpha],[\alpha]) = 0$ follows that $\alpha \in \mathrm{Deg}(h_p)$ and therefore $[\alpha] = 0$ (see (0.3.10)). So $h' = (h'_p)$ is a metric on $D = (D_p)$.

Moreover, h'_p induces a metric $\widetilde{h}_p$ on H_p by

$$\widetilde{h}_p(\xi, \eta) = h'_p(\widetilde{\psi}_p^{-1}(\xi), \widetilde{\psi}_p^{-1}(\eta)),$$

for $\xi, \eta \in H_p$. Furthermore, $H_p = TF_p$, so $\widetilde{h} = (\widetilde{h}_p)$ is a Riemannian metric on F.

Theorem 1.4.18 *The bundle $H \subseteq TM$ is a totally geodesic subbundle.*

Proof. Let $Y \in \Gamma(H)$ and $\xi \in TM_p$ be arbitrary. Then we have to show that $\nabla_\xi Y \in H_p$. As we already know that $H_p = \mathrm{Deg}(g_p)$ we demonstrate that for $\eta \in TM_p$ arbitrary

$$g_p(\nabla_\xi Y, \eta) = 0$$

holds.

Due to the fact that $TM_p = H_p \oplus \mathbb{R} \cdot B_p$, η can be written as $\eta = \zeta + \lambda \cdot B_p$, where $\zeta \in H_p$ and $\lambda \in \mathbb{R}$. Because of

$$g_p(\nabla_\xi Y, \eta) = g_p(\nabla_\xi Y, \zeta + \lambda \cdot B_p) = g_p(\nabla_\xi Y, \zeta) + \lambda g_p(\nabla_\xi Y, B_p) = 0 + \lambda g_p(\nabla_\xi Y, B_p)$$
$$= \lambda g_p(\nabla_\xi Y, B_p),$$

we only have to show that $g_p(\nabla_\xi Y, B_p) = 0$. As $Y \in \Gamma(H)$, we know that $\omega(Y) = 0$. Finally:

$$0 = \xi(\omega(Y)) = \nabla_\xi(g(Y,B)) = \underbrace{(\nabla_\xi g)(Y_p, B_p)}_{=0,\text{ because of } \nabla g = 0} + g_p(\nabla_\xi Y, B_p) + \underbrace{g_p(Y_p, \nabla_\xi B)}_{=0,\text{ as } Y_p \in H_p} = g_p(\nabla_\xi Y, B_p),$$

which shows that the bundle $H \subseteq TM$ is totally geodesic. $\square$

So it follows directly:

Corollary 1.4.19 *Every fibre N_c is a totally geodesic submanifold.*

Now it is clear that we can restrict ∇ to H and we write $\widetilde{\nabla}_X Y = \nabla_X Y$ for $X, Y \in \Gamma(H)$. Thus, if we restrict $\widetilde{\nabla}$ to F, $\widetilde{\nabla}^F$ then is a connection on $H|_F \to F$ and as ∇ is symmetric, $\widetilde{\nabla}^F$ is also symmetric on F provided that we consider $H|_F$ to be the tangent bundle of F. Now we want the connection $\widetilde{\nabla}^F$ also to be metric with respect to $\widetilde{h}$ in order to get a candidate (and therefore the only one) for the Levi-Civita connection on the tangent bundle of F. To achieve this we have to use the condition $\nabla h = 0$. Therefore, we first have to define a connection on the bundle $D = TM^*/\mathrm{Deg}(h)$. Then we can take the same approach as we took in order to construct the metric $\widetilde{h}$. In order to avoid confusion in the following we choose these terms:

- ∇ is the symmetric connection on TM;

- ∇^* is the connection on TM^*, induced by ∇:

$$(\nabla^*_\xi \alpha)(\eta) = \xi(\alpha(Y)) - \alpha_p(\nabla_\xi Y),$$

where $\xi, \eta \in TM_p$, $\alpha \in \mathcal{E}^{(1)}(M)$ and Y is a continuation of η;

- $\widetilde{\nabla}$ is the connection on H with $\widetilde{\nabla} = \nabla|_H$;

- ∇' is the connection we want to find on $D = TM^*/\mathrm{Deg}(h)$.

Theorem 1.4.20 *Let M be a four-dimensional manifold, $TM^* \to M$ the cotangent bundle, $h \in \Gamma(M; TM^* \otimes TM^*)$ a symmetric tensor field with $\mathrm{rk}(h) = 3$ and $\mathrm{ind}(h) = 0$. Let $L = \mathrm{Deg}(h) \subseteq TM^*$ and $D := TM^*/L$. Furthermore, let $h' = (h'_p)$, $h'_p : D_p \times D_p \to \mathbb{R}$ be induced by h. Finally, let ∇^* be a connection on TM^* with $\nabla^* h = 0$. Then the following holds:*

a) There is a unique connection ∇' on D with

$$\nabla'_X s' = (\nabla^*_X s)', \tag{1.6}$$

$\forall X \in \mathfrak{X}(M)$, $\forall s \in \mathcal{E}^{(1)}(M)$ and $s'(p) = \pi_p(s(p))$, where $\pi_p : TM^*_p \to D_p = TM^*_p/\mathrm{Ent}(h_p)$.

b) ∇' is metric with respect to h'.

Proof.

a) Let $p \in M$, $\xi \in TM_p$, $x : U \to V$ be a chart on M, $p \in U$ and $t \in \Gamma(U; D_{|U})$. As in the proof of (1.4.16), we now choose a bundle metric γ on TM^* and we define for $L := \mathrm{Ent}(h_p)$:

$$E := L^\perp \subseteq TM^*$$

with respect to γ. Then again $E \cong D$ by $\pi|_E : E \to D$ and we define $s := (\pi|_E)^{-1}(t)$. Therefore, we now define:
$$\nabla'_\xi t := (\nabla^*_\xi s)'.$$

Thus, if ∇' is a well defined connection on D it is already clear that is uniquely determined by (1.6).

Now let us show that ∇' is well defined and satisfies the conditions of a connection.

(i) In order to show that ∇' is well defined we have to show the following: if $u \in \Gamma(L_{|U})$, then $\nabla^*_\xi u \in \Gamma(L_{|U})$. This means that $L \subseteq TM^*$ is a parallel subbundle. Therefore, we have to show that $\nabla^*_\xi u$ is also in the degeneration space of h (remember: $L := \text{Deg}(h)$). Now, let $\alpha \in \mathcal{E}^{(1)}(M)$ be arbitrary. Then we get (because of $\nabla^*_\xi h = 0$):
$$h_p(\alpha_p, \nabla^*_\xi u) = \xi(h(\alpha, u)) - h_p(\nabla^*_\xi \alpha, u_p) = 0,$$
as $u_p \in L_p = \text{Deg}(h_p)$. Therefore ∇' is well defined.

(ii) ∇' is a connection on D: Let $X \in \mathfrak{X}(M)$, $f \in \mathcal{E}(M)$, $\omega \in \Gamma(D)$ and $\alpha := (\pi|_E)^{-1}(\omega)$, so $\alpha' = \omega$. Then we have:

(I) $\nabla'_{fX}\omega = (\nabla^*_{fX}\alpha)' = f(\nabla^*_X\alpha)' = f \cdot \nabla'_X\omega$

(II) $\nabla'_X(f\omega) = (\nabla^*_X(f\alpha))' = ((Xf) \cdot \alpha + f \cdot \nabla^*_X\alpha)' = (Xf) \cdot \alpha' + f \cdot (\nabla^*_X\alpha)'$
$= (Xf) \cdot \omega + f \cdot \nabla'_X\omega.$

b) Now let us show that ∇' is metric with respect to h', this means:
$\xi(h'(s'_1, s'_2)) = h'(\nabla'_\xi s'_1, s'_2) + h'(s'_1, \nabla'_\xi s'_2)$, for $\xi \in TM_p$ and $s_1, s_2 \in \Gamma(U; TM^*_{|U})$. Due to the definition of
$$h'_p : D_p \times D_p \to \mathbb{R}, \; h'_p(s'_1, s'_2) := h_p(s_1, s_2)$$
and $\nabla h = 0$ we have:
$$\xi(h'(s'_1, s'_2)) = \xi(h(s_1, s_2)) = h(\nabla^*_\xi s_1, s_2) + h(s_1, \nabla^*_\xi s_2) = h'((\nabla^*_\xi s_1)', s'_2) + h'(s'_1, (\nabla^*_\xi s_2)')$$
$$= h'(\nabla'_\xi s'_1, s'_2) + h'(s'_1, \nabla'_\xi s'_2).$$

And therefore ∇' is metric with respect to h'.

$\square$

Remark 1.4.21 By the bundle map $\widetilde{\psi} : D \xrightarrow{\cong} H$, ∇' induces a new connection $\widehat{\nabla}$ on H by
$$\widehat{\nabla}_X Y = \widetilde{\psi}\nabla'_X\widetilde{\psi}^{-1}(Y),$$
where $X \in \mathfrak{X}(M)$ and $Y \in \Gamma(H)$. This connection is metric with respect to $\widetilde{h}$ because of the

fact that $\nabla'_X h' = 0$, since

$$X(\widetilde{h}(Y,Z)) = X(h'(\widetilde{\psi}^{-1}(Y),\widetilde{\psi}^{-1}(Z))) = h'(\nabla'_X\widetilde{\psi}^{-1}(Y),\widetilde{\psi}^{-1}(Z)) + h'(\widetilde{\psi}^{-1}(Y),\nabla'_X\widetilde{\psi}^{-1}(Z))$$
$$= h'(\widetilde{\psi}^{-1}(\widehat{\nabla}_X Y),\widetilde{\psi}^{-1}(Z)) + h'(\widetilde{\psi}^{-1}(Y),\widetilde{\psi}^{-1}(\widehat{\nabla}_X Z)) = \widetilde{h}(\widehat{\nabla}_X Y, Z) + \widetilde{h}(Y, \widehat{\nabla}_X Z),$$

where $X \in \mathfrak{X}(M)$ and $Y, Z \in \Gamma(H)$.
On the other hand $\widetilde{\nabla}$ is symmetric on H in the sense that

$$\widetilde{T}(X,Y) = \widetilde{\nabla}_X Y - \widetilde{\nabla}_Y X - [X,Y] = 0,$$

for $X, Y \in \Gamma(H)$.

Now we will show that $\widetilde{\nabla} = \widehat{\nabla} =: \nabla^H$. If we then restrict this connection to F, we get a symmetric and metric connection on the tangent bundle $H|_F$ of F and therefore we have the candidate for the Levi-Civita connection on the leaves.

Theorem 1.4.22 *The two connections $\widetilde{\nabla}$ and $\widehat{\nabla}$ defined in the previous discussion coincide,* $\nabla^H := \widetilde{\nabla} = \widehat{\nabla}$.

Proof. In order to demonstrate the claim we first show that the map ψ (with $\psi_p : TM_p^* \to TM_p^{**} \cong TM_p$, $\alpha \mapsto h_p(\alpha,.)$) is parallel. In order to keep track we always write ∇ for any connection used in the following calculations, even if we talk about induced connections on the particular bundles (for example, $\nabla \psi$ with ∇ the induced connection on $\text{Hom}(TM^*, TM)$). Now, let $\xi \in TM_p$ as well as $\alpha, \beta \in \mathcal{E}^{(1)}(M)$ be arbitrary. Then:

$$\langle (\nabla_\xi \psi)(\alpha), \beta \rangle = \langle (\nabla_\xi(\psi(\alpha))), \beta \rangle - \langle \psi(\nabla_\xi \alpha), \beta \rangle = \langle (\nabla_\xi h(\alpha,.)), \beta \rangle - h(\nabla_\xi \alpha, \beta)$$
$$= \xi(h(\alpha, \beta)) - h(\alpha, \nabla_\xi \beta) - h(\nabla_\xi \alpha, \beta) = (\nabla_\xi h)(\alpha, \beta) = 0.$$

Furthermore, it is important that $\text{im}(\psi) = H$. Now, let $X \in \Gamma(H)$ be arbitrary. We define $\alpha := (\pi|_E)^{-1}(\widetilde{\psi}^{-1}(X))$, where $\pi|_E : E \to D$ is defined as in the proof of (1.4.20). Then $\psi(\alpha) = X$ and the following holds:

$$\widehat{\nabla}_\xi X = \widehat{\nabla}_\xi(\psi(\alpha)) = \widetilde{\psi}\nabla'_\xi\widetilde{\psi}^{-1}(\psi(\alpha)) = \widetilde{\psi}\nabla'_\xi\widetilde{\psi}^{-1}(\psi(\alpha')) = \widetilde{\psi}(\nabla'_\xi \alpha') = \widetilde{\psi}((\nabla^*_\xi \alpha)')$$
$$= \psi(\nabla^*_\xi \alpha) = \nabla_\xi(\psi(\alpha)) - \underbrace{(\nabla_\xi \psi)(\alpha)}_{=0} = \nabla_\xi X = \widetilde{\nabla}_\xi X.$$

As $X \in \Gamma(H)$ and $\xi \in TM_p$ are arbitrary, the two connections coincide. □

If we restrict the connection ∇^H to F, then $\nabla^H|_F$ is symmetric in the above sense and metric with regard to $\widetilde{h}$, and therefore the unambiguously defined Levi-Civita connection on the leave F.

If we now want to show that the curvature tensor vanishes on the leaves, we have to keep in mind that the leaves are three-dimensional. Therefore, the Ricci tensor contains all information of the curvature tensor on the leaves. As we have already seen, the leaves are totally geodesic and therefore, we can have a look at the Ricci tensor of the four-dimensional manifold M and then simply consider the components in direction of the leaves.

Theorem 1.4.23 *The leaves are flat, that means the curvature tensor vanishes, $R = 0$.*

Proof. As we have already noted we only have to show that the Ricci tensor of the whole manifold M vanishes, if restricted to the leave. According to the sixth axiom we have:

$$\operatorname{Ric}(\nabla) = 8\pi \left(T^{bb} - \frac{1}{2}\operatorname{tr}(T^b)g \right).$$

Let $\xi, \eta \in H_p$ be arbitrary. Then we get for the Ricci tensor:

$$\operatorname{Ric}(\xi, \eta) = 8\pi \left(T^{bb}(\xi, \eta) - \frac{1}{2}\operatorname{tr}(T^b) \underbrace{g_p(\xi, \eta)}_{=0} \right) = 8\pi T(\underbrace{\varphi_p(\xi)}_{=0}, \underbrace{\varphi_p(\eta)}_{=0}) = 0.$$

This shows the claim, the leaves are flat indeed. $\square$

Remark 1.4.24 a) If you replace one of the vectors ξ or η by B_p in the calculation above, the Ricci tensor still vanishes. Therefore, there are not any mixed terms of the Ricci tensor.

b) However, if you put in B_p twice you get

$$\operatorname{Ric}(B_p, B_p) = 8\pi \left(T^{bb}(B_p, B_p) - \frac{1}{2}\operatorname{tr}(T^b) \underbrace{g_p(B_p, B_p)}_{=1} \right)$$
$$= 8\pi \left(T(\varphi_p(B_p), \varphi_p(B_p)) - \frac{1}{2}T(\varphi_p(e_i), \omega^i) \right)$$
$$= 8\pi \left(T(\omega^0, \omega^0) - \frac{1}{2}T(\omega^0, \omega^0) \right) = 4\pi T(\omega^0, \omega^0),$$

where $\mathfrak{A} = (B_p, e_1, e_2, e_3)$ as well as $\mathfrak{A}^* = (\omega^0, \omega^1, \omega^2, \omega^3)$ are bases of the tangent and cotangent space respectively. This is the only component that does not vanish. Later it provides the Newtonian equation.

Note also that ω^0 does not depend on the choice of the basis but it even is $\omega = \varphi(B)$. Therefore,

$$\operatorname{Ric}(B, B) = 4\pi T(\omega, \omega)$$

holds.

1.4.5 The Ricci tensor in case of $\lambda = 0$

We want to point out that the Ricci tensor in case of $\lambda = 0$ is also symmetric. Of course, this can be seen by the calculations in (1.4.23). But it is also possible to see this directly (without using the sixth axiom). We we first have to discuss the definition of Ric in case of $\lambda = 0$. We define $\text{Ric} = (\text{Ric}_p : TM_p \times TM_p \to \mathbb{R})$ by

$$\text{Ric} := \text{tr}_{1,4}(R),$$

where $R = (R_p)$ is the curvature tensor. Thus, if $\mathfrak{A} = (e_0, e_1, e_2, e_3)$ is a basis of the tangent space and $\mathfrak{A}^* = (\omega^0, \omega^1, \omega^2, \omega^3)$ the dual basis, we define for every $p \in M$ and $\xi, \eta \in TM_p$:

$$\text{Ric}_p(\xi, \eta) = \sum_{i=0}^{3} \langle \omega^i, R_p(e_i, \xi)\eta \rangle,$$

where $\langle \_, \_ \rangle : TM_p^* \times TM_p \to \mathbb{R}$ is the natural pairing between covectors and tangent vectors. We now show that the Ricci tensor is also symmetric in case of $\lambda = 0$.

Remark 1.4.25 Let ∇ be the connection on TM as above and

$$\nabla g = 0 \text{ and } \nabla h = 0$$

with respect to the induced connections on $TM^* \otimes TM^*$ and $TM \otimes TM$. Then the dual connection ∇^* on TM^* is metric with respect to h, since the connection on $TM \otimes TM$ induced by ∇ is the same as the one induced by ∇^*.
We now have a look at $\widehat{R} : TM_p \times TM_p \times TM_p^* \times TM_p^* \to \mathbb{R}$ defined by

$$\widehat{R}_p(\xi, \alpha, \eta, \beta) := \langle \beta, R_p(\xi, \psi_p(\alpha))\eta \rangle,$$

for $\xi, \eta \in TM_p$, $\alpha, \beta \in TM_p^*$. The following holds:

Lemma 1.4.26 $\widehat{R}$ is the tensor field on $T^{(2,2)}M$ which is induced by the connection ∇^* on (TM^*, h): $\widehat{R} = \text{Rm}(\nabla^*, h)$. Thus,

$$\widehat{R}_p(\xi, \eta, \alpha, \beta) = h_p(R_p^*(\xi, \eta)\alpha, \beta),$$

where $p \in M$, $\xi, \eta \in TM_p$, $\alpha, \beta \in TM_p^*$ and $R^* \in \Gamma(M; T^{(0,2)}M \otimes \text{End}(TM^*))$ is the curvature tensor of ∇^*.

Proof. Since $\nabla^* h = 0$, we get $\nabla \psi = 0$, where ∇ is the induced connection on $\text{Hom}(TM^*, TM)$. Therefore,

$$h_p(R_p^*(\xi, \eta)\alpha, \beta) = \langle \beta, \psi_p(R_p^*(\xi, \eta)\alpha) \rangle$$

and

$$\psi_p(R_p^*(\xi, \eta)\alpha) = \psi_p(\nabla_\xi^* \nabla_Y^* \omega - \nabla_\eta^* \nabla_X^* \omega - \nabla_{[X,Y]_p}^* \omega)$$

holds, where $p \in M$, $\xi, \eta \in TM_p$, $\alpha, \beta \in TM_p^*$ and $X, Y \in \mathfrak{X}(M)$ as well as $\omega \in \mathcal{E}^{(1)}(M)$ are continuations of ξ, η and α respectively. But as $\nabla \psi = 0$, the following holds:

$$\psi_p(\nabla_\xi^* \nabla_Y^* \omega) = \nabla_\xi(\psi(\nabla_Y^* \omega)) = \nabla_\xi \nabla_Y(\psi(\omega)).$$

Therefore,
$$h_p(R_p^*(\xi,\eta)\alpha,\beta) = \langle \beta, \psi_p(R_p^*(\xi,\eta)\alpha)\rangle = \langle \beta, R_p(\xi,\eta)\psi(\alpha)\rangle = \widehat{R}_p(\xi,\eta,\alpha,\beta).$$

$\square$

Corollary 1.4.27 *The tensor $\widehat{R}$ is anti-symmetric in the last two arguments.*

Proof. This can be concluded from the fact that $\widehat{R}$ is the Riemannian curvature tensor of (∇^*, h) and that the symmetries of the Riemannian curvature tensor also hold for h degenerate (see (0.2.6)). $\square$

We now define the following contraction of $R = R(\nabla)$:
$$\operatorname{tr}_{3,4}(R) = (\operatorname{tr}_{3,4}(R_p) : TM_p \times TM_p \to \mathbb{R}); \quad \operatorname{tr}_{3,4}(R_p)(\xi,\eta) := \operatorname{spur}(R_p(\xi,\eta)).$$

Proposition 1.4.28 *Let M, g, h and ∇ be as above. Then: $\operatorname{tr}_{3,4}(R) = 0$.*

Proof. Let $p \in M$ and (e_0, e_1, e_2, e_3) a normal basis for (TM_p, g_p, h_p) with the dual basis $(\omega^0, \omega^1, \omega^2, \omega^3)$. This means that $\varphi(e_0) = \omega^0$ and $\psi(\omega^i) = e^i$ $(1 \leq i \leq 3)$. Then
$$\operatorname{tr}_{3,4}(R_p)(\xi,\eta) = \sum_{i=0}^{3} \langle \omega^i, R_p(\xi,\eta)e_i\rangle = \langle \varphi(e_0), R_p(\xi,\eta)e_0\rangle + \sum_{i=1}^{3} \langle \omega^i, R_p(\xi,\eta)\psi(\omega^i)\rangle$$
$$= g_p(R_p(\xi,\eta)e_0, e_0) + \sum_{i=1}^{3} \widehat{R}_p(\xi,\eta,\omega^i,\omega^i) = 0$$

holds because of the anti-symmetry of the Riemannian tensor in the last two arguments. $\square$

Corollary 1.4.29 *The Ricci tensor $\operatorname{Ric} := \operatorname{tr}_{1,4}(R)$ is symmetric.*

Proof. Let $p \in M$, $\xi, \eta \in TM_p$, $\mathfrak{A} = (e_0, e_1, e_2, e_3)$ be a basis of the tangent space and $\mathfrak{A}^* = (\omega^0, \omega^1, \omega^2, \omega^3)$ the dual basis. Then the following holds:
$$\operatorname{Ric}_p(\xi,\eta) = \sum_{i=0}^{3} \langle \omega^i, R_p(e_i,\xi)\eta\rangle \stackrel{(*)}{=} -\sum_{i=0}^{3} \langle \omega^i, R_p(\eta, e_i)\xi\rangle - \sum_{i=0}^{3} \langle \omega^i, R_p(\xi,\eta)e_i\rangle$$
$$= \sum_{i=0}^{3} \langle \omega^i, R_p(e_i,\eta)\xi\rangle - \underbrace{\operatorname{tr}_{3,4}(R_p)(\xi,\eta)}_{=0} = \operatorname{Ric}_p(\eta,\xi),$$

where we used the first Bianchi identity in $(*)$. $\square$

1.4.6 The global structure of M

Now we are going to introduce a new concept in order to finally show that the leaves $N_c \cong \mathbb{R}^3$ and as a consequence $M \cong \mathbb{R} \times \mathbb{R}^3$.

1.4. The case $\lambda = 0$

Definition 1.4.30 a) We call a curve $\alpha : I \to M$ *spacelike*, if there is an $\omega \in TM_p^*$ for all $\dot{\alpha}(s)$, $s \in [0, t]$, with $\psi_p(\omega) = \dot{\alpha}(s)$ and for ω

$$h_p(\omega, \omega) > 0$$

holds, for $p \in M$.

b) The manifold M is called *geodesically complete in space direction*, if every maximal spacelike geodesic is defined for all $t \in \mathbb{R}$.

Remark 1.4.31 In theorem (1.4.23) we have shown that the N_c are flat for all $c \in \mathbb{R}$. If in addition we require that M is geodesically complete in space direction the leaves N_c are geodesically complete in the Riemannian sense and thus $(N_c, \tilde{h})$ is a complete three-dimensional flat Riemannian manifold. Since $\pi_1(N_c) = (1)$, we know from Riemannian Geometry that for the N_c we have

$$(N_c, \tilde{h}) \cong \mathbb{E}^3,$$

where $\mathbb{E}^3 = (\mathbb{R}^3, g_{\text{eucl}})$ denotes the three-dimensional euclidean space (see for example [7], §8.4). By theorem (1.4.10) the manifold M has the form

$$M \cong \mathbb{R} \times F$$

in case of $\lambda = 0$. Moreover, we have shown that $F \cong \mathbb{R}^3$, altogether our manifold has the structure

$$M \cong \mathbb{R} \times \mathbb{R}^3.$$

We now want to get global coordinates for M. Therefore, let (e_1, e_2, e_3) be an orthonormal basis of TF_p and $\iota_p : \mathbb{R}^3 \to TF_p$ the corresponding coordinate isomorphism. Furthermore, let B be a complete global timelike unit vector field on M and $T : M \to \mathbb{R}$ be the fibration from (1.4.12). Let $\sigma : \mathbb{R} \to M$ be a smooth section in M with $T \circ \sigma = \text{id}_\mathbb{R}$, which means that σ is an integral curve for B. Then we get a diffeomorphism $\Phi : \mathbb{R} \times \mathbb{R}^3 \to M$ and global coordinates for M by

$$\Phi : \mathbb{R} \times \mathbb{R}^3 \to M, \ (t, x) \mapsto \exp_{\sigma(t)}(\iota_{\sigma(t)}(x)).$$

This, of course, is a diffeomorphism, since $\exp_p : TM_p \to M$ is a diffeomorphism in this case (see for instance [7], §8.4).

We now want to have a look at the liberty we have for the coordinate system of M. In every $\sigma(t)$, where σ is an integral curve for B, we have to choose $(e_1(t), e_2(t), e_3(t))$ (orthonormal) in differentiable dependence on t. Therefore, the only choices we have are the following:

Let $c \in \mathbb{R}$, $b : \mathbb{R} \to \mathbb{R}^3$ and $S : \mathbb{R} \to O(3)$. Then every possible coordinate transformation is given by

$$\tau : \mathbb{R} \times \mathbb{R}^3 \to \mathbb{R} \times \mathbb{R}^3, \ (t, x) \mapsto (\pm t + c, S(t)x + b(t)).$$

In (1.4.10) we have already found a diffeomorphism $\Psi : \mathbb{R} \times F \to M$ depending on B. Note that the coordinates given by this diffeomorphism are not always of the form mentioned above.

Having all this in mind we now can formulate the following

Theorem 1.4.32 *Let M be a four-dimensional connected smooth manifold, $g \in \Gamma(M; T^{(0,2)}M)$, $h \in \Gamma(M; T^{(2,0)}M)$, $\lambda = 0$, and ∇ a symmetric connection on M. These mathematical objects have to satisfy the following axioms of the frame theory:*

1. *At every point $p \in M$ there is a timelike tangent vector $\xi \in TM_p$, and h_p is positive definite on the annihilator of ξ,*

$$h_p|_{Ann(\xi) \times Ann(\xi)} > 0.$$

2. *For the transitions from the tangent space to the cotangent space and vice versa induced by g and h*

$$\varphi_p \circ \psi_p = -\lambda \operatorname{id}_{TM_p^*}$$

holds for all $p \in M$.

3. *Both g and h have to be parallel with regard to the connection ∇,*

$$\nabla g = \nabla h = 0.$$

4.
$$\operatorname{Ric}(\nabla) = 8\pi \left(T^{bb} - \frac{1}{2} \operatorname{tr}(T^b) g \right).$$

If M is simply connected, then there is a global timelike unit vector field on M. If one of these vector fields is complete and, in addition, M geodesically complete in space direction, then M is a Newtonian space, which means that

$$M \cong \mathbb{E}^4.$$

Remark 1.4.33 We want to point out once more that if we do not require the manifold M to be simply connected and geodesically complete in space direction or the timelike unit vector field to be complete, we locally still get what we want. Then the manifold M is of the form $I \times U$, where $I \subseteq \mathbb{R}$ is an open interval and U a flat three dimensional manifold.

1.4.7 The Newtonian equations

The components of the connection 1.4.34 In case of $\lambda = 0$ we have seen that the manifold has the desired Newtonian structure. Next we are interested in the Newtonian equations which emerge from Einstein's field equations in case of $\lambda = 0$.

So let M, g, h, and ∇ be as above and (t, x^1, x^2, x^3) a coordinate system for M. We already know from the discussion above that $\nabla_X Y \in \Gamma(H)$ for all $X, Y \in \mathfrak{X}(M)$ (see (1.4.6) and (1.4.18)). As ∇ is flat, we also know that $\nabla_{\partial_i}\partial_j = 0$ for $1 \leq i \leq 3$, which means that ∇ is defined by ∇B, $\left(B = \frac{\partial}{\partial t}\right)$. We therefore define

$$\nabla_B B = \Gamma^i_{00}\partial_i =: f^i \partial_i$$

and

$$\nabla_{\partial_i} B = \Gamma^j_{i0}\partial_j =: v^j_i \partial_j$$

for $1 \leq i, j \leq 3$. Due to $\nabla h = 0$ we then get

$$\begin{aligned}
v^j_i &= \langle dx^j, \nabla_{\partial_i} B \rangle = \langle dx^j, \nabla_B \partial_i \rangle = \langle -\nabla_B dx^j, \partial_i \rangle = -\langle \nabla_B dx^j, \psi(dx^i) \rangle \\
&= -h(\nabla_B dx^j, dx^i) = h(dx^j, \nabla_B dx^i) = \langle \nabla_B dx^i, \psi(dx^j) \rangle = \langle \nabla_B dx^i, \partial_j \rangle \\
&= -\langle dx^i, \nabla_B \partial_j \rangle = -\langle dx^i, \nabla_{\partial_j} B \rangle = -v^i_j.
\end{aligned}$$

Therefore, the only components of ∇ which do not vanish are

$$\Gamma^i_{00} = f^i, \quad i = 1, 2, 3;$$
$$\Gamma^j_{0i} = \Gamma^j_{i0} = v^j_i = -v^i_j, \quad i, j = 1, 2, 3.$$

Particularly,

$$\Gamma^i_{0i} = 0,$$

which means that ∇ is determined by the six functions f^1, f^2, f^3 and v^3_2, v^1_3, v^2_1.

Remark 1.4.35 The functions f^1, f^2, f^3 and v^3_2, v^1_3, v^2_1 are not determined any further by the axioms of the frame theory used up to this moment. With the help of the last three axioms we can find out more about these functions.

Now we try to derive the Newtonian equations with the help of the last three axioms. As we will soon see, in most cases the so called "Quasi-Newtonian" equations appear, not genuine Newtonian ones. In the following chapters we therefore also have to discuss the question when genuine Newtonian models appear.

The components of the Ricci tensor 1.4.36 First we talk about the components of the Ricci tensor which do not vanish and we use the last axiom of the frame theory. By (1.4.24) we already know that

$$\text{Ric}(B, B) = 4\pi T(\omega, \omega).$$

Remember that $\omega \in \mathcal{E}^{(1)}(M)$ is the inambiguously defined 1-form on M (except for the sign) with $\ker(\omega) = H$ and $\widetilde{g}(\omega, \omega) = 1$, where $\widetilde{g}$ is the metric on $L = \text{Deg}(h) \subseteq TM^*$ induced by g. Therefore, $\rho := T(\omega, \omega)$ does not depend on coordinates and we can call it *mass density* on M.

As furthermore

$$\mathrm{Ric}_p(B_p, B_p) = \mathrm{Ric}_p(e_0, e_0) = \sum_{i=0}^{3} \mathrm{Ric}_p(e_i, e_i) = \mathrm{tr}_g(\mathrm{Ric})_p,$$

where (B_p, e_1, e_2, e_3) is an orthonormal basis of TM_p, we define

$$S := \mathrm{Ric}(B, B),$$

where S is the scalar curvature of (M, g, h, ∇). This means that we have

$$S = 4\pi\rho.$$

Now let $X \in \Gamma(H)$, (B_p, e_1, e_2, e_3) be an orthonormal basis of TM_p and $(\omega, \omega^1, \omega^2, \omega^3)$ dual with $\varphi_p(B_p) = \omega$. Then

$$\mathrm{Ric}(B, X)_p = \sum_{i=1}^{3} \langle \omega^i, R_p(e_i, B_p) X_p \rangle$$

holds, as $\omega^0 = \omega = \varphi_p(B_p)$ and $R_p(\xi, \eta)\zeta \in H_p$ for all $\xi, \eta, \zeta \in TM_p$. In our coordinates (t, x^1, x^2, x^3) with $B = \frac{\partial}{\partial t}$ and $e_i = \partial_i$ we then get:

$$R(e_i, B)e_j = \nabla_{\partial_i} \nabla_B \partial_j - \nabla_B \underbrace{\nabla_{\partial_i} \partial_j}_{=0} - \nabla_{\underbrace{[B, \partial_i]}_{=0}} \partial_j = -\nabla_{\partial_i} \nabla_{\partial_j} B.$$

Therefore, we get for the Ricci tensor:

$$\mathrm{Ric}(B, \partial_j) = -\sum_{i=1}^{3} \langle \omega^i, -\nabla_{\partial_i} \nabla_{\partial_j} B \rangle = -\mathrm{tr}_H(\mathrm{Hess}_H(B))_j = -\Delta^H(B)_j,$$

where the index H always means that we only take the components in direction of H. As we already know that all components of the Ricci tensor vanish (except $\mathrm{Ric}(B, B)$), we therefore get the equation

$$\Delta^H B = 0.$$

But this means that we now know all components of the Ricci tensor and that axiom six $(\mathrm{Ric}(\nabla) = 8\pi \left(T^{bb} - \frac{1}{2} \mathrm{tr}(T^b) g \right))$ is equivalent to the following three conditions in case of $\lambda = 0$:

- All Frobenius leaves are flat, $R|_{H \times H \times H} = 0$;
- $S = 4\pi\rho$;
- $\Delta^H B = 0$.

Now we want to discuss the last two equations in coordinates. We get:

$$S(p) = \mathrm{Ric}(B,B) = \sum_{i=1}^{3}\langle \omega^i, R(e_i,B)B\rangle = \sum_{i=1}^{3}\langle \omega^i, \nabla_{\partial_i}\nabla_B B - \nabla_B \nabla_{\partial_i} B\rangle$$

$$= \sum_{i=1}^{3}\langle \omega^i, \nabla_{\partial_i}(f^j \partial_j) - \nabla_B(v_i^j \partial_j)\rangle = \sum_{i=1}^{3}\langle \omega^i, (D_i f^j)\partial_j - B v_i^j \partial_j - v_i^j \nabla_B \partial_j\rangle$$

$$= \sum_{i=1}^{3}\langle \omega^i, (D_i f^k - D_t v_i^k - v_i^j v_j^k)\partial_k\rangle = \sum_{i=1}^{3}(D_i f^i - D_t \underbrace{v_i^i}_{=0} - v_i^j v_j^i) = \mathrm{div}(f) - 2\|v\|^2$$

(in cartesian coordinates). We therefore get the equation

$$\mathrm{div}(f) - 2\|v\|^2 = 4\pi\rho. \qquad (1.7)$$

The second condition in coordinates equals

$$\langle \omega^j, \nabla^H B\rangle = \sum_{i=1}^{3}\langle \omega^i, \nabla_{\partial_i}\nabla_{\partial_j} B\rangle = \sum_{i=1}^{3}\langle \omega^i, \nabla_{\partial_i}(v_j^k \partial_k)\rangle = \sum_{i=1}^{3}\langle dx^i, D_i v_j^k \partial_k\rangle = \sum_{i=1}^{3} D_i v_j^i.$$

If we now define $v_i := \epsilon_{ij}^k v_k^j$, we get $D_i v_j^i = \mathrm{rot}(v)_j$ and therefore, $\Delta^H B = 0$ is equivalent to

$$\mathrm{rot}(v) = 0, \qquad (1.8)$$

where $\epsilon_{ij}^k = 1$ if (i,j,k) is an even permutation of $(1,2,3)$, $\epsilon_{ij}^k = -1$, if (i,j,k) is an odd permutation of $(1,2,3)$ and $\epsilon_{ij}^k = 0$, otherwise.

The symmetry of the tensor $R^\#$: the fifth axiom 1.4.37 Now we have to discuss axiom five, which contains the symmetry of the tensor $R^\#$:

$$R_p^\#(\xi, \lambda, \eta, \mu) = R_p^\#(\eta, \mu, \xi, \lambda). \qquad (1.9)$$

First, we have to remember that we had:

$$R_p^\# : TM_p \times TM_p^* \times TM_p \times TM_p^* \to \mathbb{R}, \quad R_p^\#(\xi, \lambda, \eta, \mu) := \langle \mu, R_p(\xi, \psi_p(\lambda))\eta\rangle,$$

where $R_p : TM_p \times TM_p \times TM_p \to TM_p$ is the curvature tensor and $\psi_p : TM_p^* \to TM_p$ is given by

$$\langle \lambda, \psi_p(\mu)\rangle = h_p(\lambda, \mu), \qquad (1.10)$$

for all $\lambda, \mu \in TM_p^*$. Furthermore, we have to remember the bundle metric $\widetilde{h}_p : H_p \times H_p \to \mathbb{R}$ defined on $H_p = \mathrm{Deg}(g_p)$ by

$$\widetilde{h}_p(\psi_p(\lambda), \psi_p(\mu)) = h_p(\lambda, \mu).$$

Thus, due to (1.10) this means that

$$\tilde{h}_p(\psi_p(\lambda), \psi_p(\mu)) = \langle \lambda, \psi_p(\mu) \rangle,$$

and, since $\operatorname{im}(\psi_p) = H_p$,

$$\tilde{h}_p(\psi_p(\lambda), \eta) = \langle \lambda, \eta \rangle$$

also holds for all $\eta \in H_p$. Therefore, it follows that

$$R_p^{\#}(\xi, \lambda, \eta, \mu) = \tilde{h}_p(R_p(\xi, \psi_p(\lambda))\eta, \psi_p(\mu)),$$

since we already know that $R_p(\xi, \eta) : TM_p \to TM_p$ always has its values in H_p. But this means that for $\xi, \eta \in H_p$

$$R_p^{\#}(\xi, \lambda, \eta, \mu) = \tilde{h}_p(R_p(\xi, \psi_p(\lambda))\eta, \psi_p(\mu)) = \operatorname{Rm}_p^{N_p}(\xi, \psi_p(\lambda), \eta, \psi_p(\mu))$$

holds, where Rm^{N_p} is the Riemannian curvature tensor of the Riemannian manifold $(N_p, \tilde{h}|_{N_p})$. Therefore the symmetry condition (1.9) already holds for $\xi, \eta \in H_p$.

Thus, condition (1.9) only provides new information in the two following cases:

a) $\xi = \eta = B_p$

b) $\xi \in H_p, \eta = B_p$.

So we again choose a coordinate system (t, x) and, of course,

$$\nabla_B B = f^i \partial_i$$

and

$$\nabla_{\partial_i} B = v_i^j \partial_j$$

hold for $1 \leq i, j \leq 3$, where $B = \partial_t$. Furthermore, we know that $\nabla_{\partial_i} \partial_j = 0$. And we know from the calculations about the scalar curvature in (1.4.36) that

$$R(B, \partial_i)B = (Bv_i^j - D_i f^j + v_i^k v_k^j)\partial_j.$$

Now we can define for $\lambda, \mu \in TM_p$, $\lambda = dx^i$ and $\mu = dx^j$ so that $\psi(\lambda) = \partial_i$ and $\psi(\mu) = \partial_j$ (remember that the condition always holds for $\lambda = \omega = \varphi(B)$ or $\mu = \omega = \varphi(B)$, as both sides already vanish).

But this means that

$$R^{\#}(B, dx^i, B, dx^j) = Bv_i^j - D_i f^j + v_i^k v_k^j.$$

Then condition (1.9) results in

$$0 = R^\#(B, dx^i, B, dx^j) - R^\#(B, dx^j, B, dx^i) = (Bv_i^j - Bv_j^i) - (D_i f^j - D_j f^i) + (v_i^k v_k^j - v_j^k v_k^i)$$
$$= 2Bv_i^j - (D_i f^j - D_j f^i) + \underbrace{(v_i^k v_k^j - (-v_i^k)(-v_k^j))}_{=0},$$

for all $1 \le i, j \le 3$, due to the condition $v_j^i = -v_i^j$. If we again define $v_k = \epsilon_{jk}^i v_i^j$ and $v = (v_1, v_2, v_3)$, this results in

$$2\partial_t v + \text{rot}(f) = 0. \tag{1.11}$$

Now let $\xi := \partial_i$, $\lambda := dx^j$ and $\mu := dx^k$ for $1 \le i, j, k \le 3$. Again we know from (1.4.36) that

$$R(B, \partial_k)\partial_i = D_k v_i^j \partial_j.$$

Thus,

$$R^\#(B, dx^k, \partial_i, dx^j) = D_k v_i^j.$$

Furthermore, we calculate

$$R(\partial_i, \partial_j)B = \nabla_{\partial_i}\nabla_{\partial_j}B - \nabla_{\partial_j}\nabla_{\partial_i}B = \nabla_{\partial_i}(v_j^k \partial_k) - \nabla_{\partial_j}(v_i^k \partial_k) = (D_i v_j^k - D_j v_i^k)\partial_k.$$

Thus,

$$R^\#(\partial_i, dx^j, B, dx^k) = D_i v_j^k - D_j v_i^k.$$

This means that condition (1.9) results in

$$D_i v_j^k + D_j v_k^i + d_k v_i^j = 0$$

for all $1 \le i, j, k \le 3$. But this condition always holds if two of the indices are the same. And if it holds for one permutation of $(1, 2, 3)$, then also for all the other five. So if we define $i = 1$, $j = 2$ and $k = 3$, this results in

$$\text{div}(v) = 0. \tag{1.12}$$

All in all, this discussion shows that the symmetry condition (1.9) for $R^\#$ in coordinates is equivalent to the two equations

$$2D_t v + \text{rot}(f) = 0$$
$$\text{div}(v) = 0.$$

The divergence of the matter tensor T: the fourth axiom 1.4.38 Now we discuss the fourth axiom of the frame theory:

$$\text{div}(T) = 0.$$

1.4. The case $\lambda = 0$

We again use our coordinate system (t, x) and define:

$$\rho := T(dt, dt), \quad j^k := T(dt, dx^k), \quad S^{kl} := T(dx^k, dx^l),$$

for all $1 \leq k, l \leq 3$ and as before

$$\nabla_B B = f^i \partial_i, \quad \nabla_{\partial_i} B = \nabla_B \partial_i = v_i^k \partial_k, \quad \nabla_{\partial_i} \partial_k = 0,$$

for all $1 \leq i, k \leq 3$. Then we first have for $1 \leq i, k, l \leq 3$:

$$\langle \nabla_{\partial_i} dx^k, \partial_l \rangle = D_i \underbrace{\langle dx^k, \partial_l \rangle}_{\delta_l^k} - \langle dx^k, \underbrace{\nabla_{\partial_i} \partial_l}_{=0} \rangle = 0$$

$$\langle \nabla_{\partial_i} dx^k, B \rangle = D_i \underbrace{\langle dx^k, B \rangle}_{\delta_0^k} - \langle dx^k, \nabla_{\partial_i} B \rangle = -v_i^k.$$

Thus,

$$\nabla_{\partial_i} dx^k = -v_i^k dt.$$

Furthermore, it holds for $1 \leq i, l \leq 3$:

$$\langle \nabla_{\partial_i} dt, \partial_l \rangle = -\langle dt, \underbrace{\nabla_{\partial_i} \partial_l}_{=0} \rangle = 0$$

$$\langle \nabla_{\partial_i} dt, B \rangle = -\langle dt, \nabla_{\partial_i} B \rangle = -\langle dt, v_i^k \partial_k \rangle = 0.$$

Thus,

$$\nabla_{\partial_i} dt = 0.$$

Moreover, for $1 \leq l \leq 3$:

$$\langle \nabla_B dt, \partial_l \rangle = -\langle dt \nabla_B \partial_l \rangle = -\langle dt, v_l^k \partial_k \rangle = 0$$

$$\langle \nabla_B dt, B \rangle = -\langle dt, \nabla_B B \rangle = -\langle dt, f^k \partial_k \rangle = 0.$$

Thus,

$$\nabla_B dt = 0.$$

Finally, for $1 \leq i, l \leq 3$:

$$\langle \nabla_B dx^i, \partial_l \rangle = -\langle dx^i \nabla_B \partial_l \rangle = -\langle dx^i, v_l^k \partial_k \rangle = -v_l^i$$

$$\langle \nabla_B dx^i, B \rangle = -\langle dx^i, \nabla_B B \rangle = -\langle dx^i, f^k \partial_k \rangle = -f^i.$$

Thus,
$$\nabla_B dx^i = -v_l^i dx^l - f^i dt.$$

Now we are able to calculate the components of the divergence of T. Remember that we had
$$\text{div}(T) = \text{tr}_{(1,3)}(\nabla T).$$

We therefore get:

$$\langle dt, \text{div}(T) \rangle = \sum_{i=0}^{3} \nabla_{\partial_i} T(dx^i, dt) = \nabla_B T(dt, dt) + \sum_{i=1}^{3} \nabla_{\partial_i} T(dx^i, dt)$$
$$= B(T(dt, dt)) - T(\nabla_B dt, dt) - T(dt, \nabla_B dt)$$
$$+ \sum_{i=1}^{3} (\partial_i(T(dx^i, dt)) - T(\nabla_{\partial_i} dx^i, dt) - T(dx^i, \nabla_{\partial_i} dt))$$
$$= D_t \rho + \sum_{i=1}^{3} (D_i j^i - T(\underbrace{v_i^i}_{=0} dt, dt)) = D_t \rho + \text{div}(j).$$

Furthermore, for $1 \leq l \leq 3$:

$$\langle dx^l, \text{div}(T) \rangle = \sum_{i=0}^{3} \nabla_{\partial_i} T(dx^i, dx^l) = \nabla_B T(dt, dx^l) + \sum_{i=1}^{3} \nabla_{\partial_i} T(dx^i, dx^l)$$
$$= B(T(dt, dx^l)) - T(\nabla_B dt, dx^l) - T(dt, \nabla_B dx^l)$$
$$+ \sum_{i=1}^{3} (\partial_i(T(dx^i, dx^l)) - T(\nabla_{\partial_i} dx^i, dx^l) - T(dx^i, \nabla_{\partial_i} dx^l))$$
$$= D_t(j^l) - T(dt, -v_i^l dx^i - f^l dt) + \sum_{i=1}^{3}(D_i(S^{il}) - T(\underbrace{v_i^i}_{=0} dt, dx^l) - T(dx^i, -v_i^l dt))$$
$$= D_t(j^l) + v_i^l T(dt, dx^i) + f^i T(dt, dt) + D_i(S^{il}) + v_i^l T(dx^i, dt)$$
$$= D_t(j^l) + v_i^l j^i + f^l \rho + D_i(S^{il}) + v_i^l j^i = D_t(j^l) + D_i(S^{il}) + 2v_i^l j^i + \rho f^l.$$

If we again define $v_k = \epsilon^i_{jk} v^j_i$ and $v = (v_1, v_2, v_3)$ we see that the condition $\text{div}(T) = 0$ is equivalent to the two equations

$$D_t \rho + \text{div}(j) = 0$$
$$D_t(j) + \text{div}(S) = 2j \times v - \rho f.$$

Remark 1.4.39 If you remember the discussion about a perfect fluid (see (0.3.16)) you can realize that the two equations mentioned above contain the Euler equations as a special case. First, we require v to disappear, $v = 0$, and ρ to be constant (but not zero). Furthermore, we define $S := pE_3$, $j := \rho V$ and $-G = f$. Then we get:

$$D_t j + \operatorname{div}(S) = j \times v - \rho f \Leftrightarrow D_t(\rho V) + \operatorname{grad}(p) = \rho G \Leftrightarrow \rho D_t V = -\operatorname{grad}(p) + \rho G,$$

since ρ does not depend on t and since the following holds:

$$\operatorname{div}(S)^j = \frac{\partial S^{ij}}{\partial x^i} = \frac{\partial(p \cdot \delta^{ij})}{\partial x^i} = \delta^{ij}\frac{\partial p}{\partial x^i} = \frac{\partial p}{\partial x^j} = \operatorname{grad}(p)^j.$$

The other equation results in

$$0 = \operatorname{div}(\rho V) = \rho \operatorname{div}(V) \Rightarrow \operatorname{div}(V) = 0,$$

since ρ is constant and not zero. But this means that the equations which emerge from the axiom $\operatorname{div}(T) = 0$ contain the Euler equations as a special case.

Law of motion 1.4.40 Finally, we study the local geodesic equation in case of $\lambda = 0$ for the special coordinates named above. In general, these equations read:

$$\ddot{x}^k + \Gamma^k_{ij}(x)\dot{x}^i\dot{x}^j = 0.$$

Then we get for $k = 1$:

$$\ddot{x}^1 = -\Gamma^1_{00}\dot{x}^0\dot{x}^0 + 2\Gamma^1_{02}\dot{x}^0\dot{x}^2 + 2\Gamma^1_{03}\dot{x}^0\dot{x}^3 = f^1\dot{x}^0\dot{x}^0 - 2v_{12}\dot{x}^0\dot{x}^2 + 2v_{31}\dot{x}^0\dot{x}^3$$
$$= f^1 - 2v_{12}\dot{x}^2 + 2v_{31}\dot{x}^3,$$

where (without restriction)

$$\ddot{x}^0 = 0 \Rightarrow (\dot{x})^0 = 1 \Rightarrow x^0(t) = t$$

holds. In exactly the same way we have for $k = 2$ and $k = 3$:

$$\ddot{x}^2 = f^2 + 2v_{12}\dot{x}^1 - 2v_{23}\dot{x}^3$$
$$\ddot{x}^3 = f^3 - 2v_{31}\dot{x}^1 + 2v_{23}\dot{x}^2.$$

Altogether this means:

$$\ddot{x} = f - 2\dot{x} \times v, \tag{1.13}$$

where $\times$ is the cross product in $\mathbb{R}^3$.

Remark 1.4.41 (i) In the previous discussion some notations appear that may surprise some readers. We consider v and f to be time dependent vector fields on $\mathbb{R}^3$ in our special coordinates. Only then it makes sense to talk about $\operatorname{rot}(f)$, $\operatorname{rot}(v)$ and the cross product in $\mathbb{R}^3$.

(ii) If $\operatorname{rot}(f) = 0$, f can be written as the gradient of a map u, $f = \operatorname{grad}(u)$. This is the case if and only if $\partial_0(v) = 0$ (see (1.11)). Therefore, this is also necessary if we want to get a Newtonian model. But we first have to define what this exactly means in our context. This is one task of the next chapter.

We now summarize the results of the first chapter:

Theorem 1.4.42 *Let M be a four-dimensional connected smooth manifold, $\lambda \in \mathbb{R}_0^+$, $g \in \Gamma(M; T^{(0,2)}M)$, $h \in \Gamma(M; T^{(2,0)}M)$, $T \in \Gamma(M; T^{(2,0)}M)$, and ∇ a symmetric connection on M. These mathematical objects have to satisfy the six axioms of the frame theory. Then the following holds:*

a) *For $\lambda > 0$, $(M, (-\lambda g))$ is a spacetime, ∇ the Levi-Civita connection of $(-\lambda g)$ and $(-\lambda g)$ satisfies Einstein's field equations for the given T.*

b) *For $\lambda = 0$ the following holds: If M is simply connected, then there is a global timelike unit vector field on M. If one of these vector fields is complete and, in addition, M geodesically complete in space direction, M is a Newtonian space, which means that*

$$M \cong \mathbb{E}^4.$$

Furthermore, there are global coordinates on M with regard to which we get the quasi-Newtonian equations

(i) $\operatorname{div}(f) - 2v^2 = 4\pi\rho$,

(ii) $\operatorname{rot}(v) = 0$,

(iii) $\operatorname{rot}(f) + 2\,\dot{v} = 0$,

(iv) $\operatorname{div}(v) = 0$,

(v) $D_t \rho + \operatorname{div}(j) = 0$,

(vi) $D_t(j) + \operatorname{div}(S) = 2j \times v - \rho f$,

(vii) $\ddot{x} = f - 2\dot{x} \times v$,

where $f^i = \Gamma_{00}^i$, for $i = 1, 2, 3$, $v_{ij} = \Gamma_{0i}^k$, $-v_{ik} = \Gamma_{0k}^i$, for $i, k \in \{1, 2, 3\}$, $T^{00} = \rho$, $T^{0k} = j^k$ and $T^{ik} = S^{ik}$, for $i, k \in \{1, 2, 3\}$.

Remark 1.4.43 Of course, we are interested in genuine Newtonian equations. For $v = 0$ the quasi-Newtonian equations result in

(i) $\operatorname{div}(f) = 4\pi\rho$,

(ii) $\operatorname{rot}(f) = 0$,

(iii) $D_t \rho + \operatorname{div}(j) = 0$

(iv) $D_t(j) + \operatorname{div}(S) = -\rho f$

(v) $\ddot{x} = f$,

1.4. The case $\lambda = 0$

which means that we get the Newtonian law of motion and as mentioned above, f can be written as the gradient of a map u, $f = \text{grad}(u)$. Therefore, the law of gravitation

$$\Delta u = 4\pi\rho$$

also holds in this case. Furthermore, we have seen in (1.4.39) that the equations in (iii) and (iv) contain the Euler equations for a perfect fluid as a special case. Thus, it is very important to discuss the question, when $v = 0$. This is the task of the third chapter.

Chapter 2

The Newtonian limit: definition and existence

In this chapter we first want to define the concept of the Newtonian limit. After that we want to discuss some examples of limits of solutions of Einstein's field equations. In the third part we discuss conditions which make sure that there is a Newtonian or a quasi-Newtonian limit. Finally, we discuss the limit of spherically symmetric and static solutions.

2.1 Definition of the Newtonian limit

In order to define a Newtonian limit we first have to study how classical Newtonian systems can be imbedded in the frame theory. Only then it really makes sense to talk about a Newtonian limit. For instance, in case of a classical Newtonian system we certainly do not have an arbitrary four-dimensional manifold but the space $\mathbb{R} \times \mathbb{R}^3$. Therefore, we now have to consider the following.

Definition 2.1.1   a) A tuple $(M, g, h, \nabla, T, \lambda)$ which meets the conditions of the frame theory is called an *Ehlers spacetime*.

b) We call $(M, g(\lambda), h(\lambda), \nabla(\lambda), T(\lambda), \lambda)$ a by $\lambda \in (0, 1]$ parametrized *family of Ehlers spacetimes*, if $(M, g(\lambda), h(\lambda), \nabla(\lambda), T(\lambda), \lambda)$ satisfies the frame theory for every $\lambda \in (0, 1]$ fixed.

Definition 2.1.2 A *classical Newtonian system* is a tuple $(\mathbb{R} \times \mathbb{R}^3, \rho, u)$ with smooth maps $\rho : \mathbb{R} \times \mathbb{R}^3 \to \mathbb{R}_+$ and $u : \mathbb{R} \times \mathbb{R}^3 \to \mathbb{R}$ so that the following holds:

(i) $\Delta u = \frac{\partial^2 u}{(\partial x^1)^2} + \frac{\partial^2 u}{(\partial x^2)^2} + \frac{\partial^2 u}{(\partial x^3)^2} = 4\pi\rho$,

(ii) $u(t, x) \to 0$ for $|x| \to \infty$.

Definition 2.1.3 We define the *imbedding of a classical Newtonian system in the category of all Ehlers spacetimes* as the functor which relates a classical Newtonian system $(\mathbb{R} \times \mathbb{R}^3, \rho, u)$ to an Ehlers spacetime $(M, g, h, \nabla, T, \lambda)$ so that (with respect to the natural coordinate system (t, x)):

(i) $M = \mathbb{R} \times \mathbb{R}^3$,

(ii) $(g_{ij}) = \mathrm{diag}(1,0,0,0)$,

(iii) $(h^{ij}) = \mathrm{diag}(0,1,1,1)$,

(iv) with $f := \mathrm{grad}(u)$ we require: for $i=1,2,3$, $\Gamma^i_{00} = f^i$, as well as $\Gamma^k_{ij} = 0$ otherwise,

(v) $T^{00} = \rho$, $T^{ij} = 0$ otherwise,

(vi) $\lambda = 0$.

Remark 2.1.4 Of course we have to discuss if the resulting tuple $(M, g, h, \nabla, T, \lambda)$ of definition (2.1.3) is an Ehlers spacetime. But the definition is suitable to the results of section (1.4.7). For instance, in (1.4.36) we found that for $\lambda = 0$ the sixth axiom

$$\mathrm{Ric}(\nabla) = 8\pi(T^{bb} - \frac{1}{2}\mathrm{tr}(T^b)g)$$

is equivalent to the three conditions (in coordinates):

- All Frobenius leaves are flat;
- $\mathrm{div}(f) - 2||v||^2 = 4\pi\rho$;
- $\mathrm{rot}(v) = 0$.

But these conditions are satisfied, of course, since $\mathbb{R}^3$ is flat, $v = 0$, thus $\mathrm{rot}(v) = 0$ and

$$\mathrm{div}(f) - ||v||^2 = \mathrm{div}(f) = \mathrm{div}(\mathrm{grad}(u)) = \Delta u = 4\pi\rho$$

holds due to the fact that $(\mathbb{R} \times \mathbb{R}^3, \rho, u)$ is a classical Newtonian system.

In exactly the same way we now define a quasi-Newtonian system and its imbedding in the category of all Ehlers spacetimes:

Definition 2.1.5 A *quasi-Newtonian system* is a tuple $(\mathbb{R} \times \mathbb{R}^3, \rho, f, v)$ with time-dependent vector fields v and f on $\mathbb{R}^3$, a map $\rho : \mathbb{R} \times \mathbb{R}^3 \to \mathbb{R}_+$ for which the following holds:

(i) $\mathrm{div}(f) - 2||v||^2 = 4\pi\rho$,

(ii) $\mathrm{rot}(v) = 0$,

(iii) $\mathrm{rot}(f) + 2D_t v = 0$,

(iv) $\mathrm{div}(v) = 0$.

Definition 2.1.6 We define the *imbedding of a quasi-Newtonian system in the category of all Ehlers spacetimes* as the functor which relates a quasi-Newtonian system $(\mathbb{R} \times \mathbb{R}^3, \rho, f, v)$ to an Ehlers spacetime $(M, g, h, \nabla, T, \lambda)$ so that:

(i) $M \subseteq \mathbb{R} \times \mathbb{R}^3$,

(ii) $(g_{ij}) = \text{diag}(1, 0, 0, 0)$,

(iii) $(h^{ij}) = \text{diag}(0, 1, 1, 1)$,

(iv) for $i = 1, 2, 3$, $\Gamma^i_{00} = f^i$, as well as $\Gamma^j_{0i} = v_{ij}$, $\Gamma^i_{0j} = -v_{ij}$, for $i, j \in \{1, 2, 3\}$, $\Gamma^k_{ij} = 0$ otherwise,

(v) $T^{00} = \rho$ and $T^{ij} = 0$ otherwise,

(vi) $\lambda = 0$.

Remark 2.1.7 As we saw in the chapter about the frame theory, an Ehlers spacetime which is related to a quasi-Newtonian system fulfils the frame theory for $\lambda = 0$, whereas an Ehlers spacetime which is related to a classical Newtonian system additionally meets the condition $v = 0$ and it is $M = \mathbb{R} \times \mathbb{R}^3$. Now it also makes sense to talk about a Newtonian or quasi-Newtonian limit of a family of Ehlers spacetimes:

Definition 2.1.8 For $\lambda \in (0, 1]$ let $(M, g(\lambda), h(\lambda), \nabla(\lambda), T(\lambda), \lambda)$ be a parametrized family of Ehlers spacetimes on a four-dimensional connected differentiable manifold M.

a) We say that this family of Ehlers spacetimes *has a quasi-Newtonian limit*, if

(i) all tensor fields have limits in every point $p \in M$ for $\lambda \to 0$. The metrics $g(\lambda)$ and $h(\lambda)$ converge in the $\mathcal{C}^2$-norm towards the limit metrics $g(0)$ and $h(0)$, the connections converge in the $\mathcal{C}^1$-norm towards the limit connection $\nabla(0)$;

(ii) the limits $(g(0), h(0), \nabla(0), T(0))$ fulfil the frame theory for $\lambda = 0$.

b) Such a family of Ehlers spacetimes *has a Newtonian limit*, if

(i) the family has a quasi-Newtonian limit;

(ii) the manifold is simply connected and geodesically complete in space direction and furthermore, there is a **complete** global timelike unit vector field;

(iii) there are Newtonian coordinates so that $v = 0$ in these coordinates.

Remark 2.1.9   a) Looking at condition (i) in the definition of a quasi-Newtonian limit it is worth noting that the connection has to converge in a $\mathcal{C}^1$-sense. This guarantees that the limit of the curvature tensor and the curvature tensor of the limit connection coincide.

b) As we saw in the first chapter the conditions (ii) and (iii) of definition (2.1.8) really guarantee that a family of Ehlers spacetimes not only has a quasi-Newtonian but also a genuine Newtonian limit. The condition (ii) makes sure that the manifold in case of $\lambda = 0$ really has the form $M = \mathbb{R} \times \mathbb{R}^3$.
Additionally, the third condition causes that the quasi-Newtonian equations transform

into the genuine Newtonian ones. Since, if $v = 0$, equation (1.11) equals $\operatorname{rot}(f) = 0$ and — as we already commented in the chapter about the frame theory — f can be written as a gradient of a map u, $f = \operatorname{grad}(u)$. Thus, the genuine Newtonian law of gravitation emerges from the equation (1.7). If we consider (1.13) in addition, we can see that it turns out to be the Newtonian law of motion.

2.2 Extension of spacetimes

All in all, it is our aim to find out when a solution of Einstein's field equations has a Newtonian limit. Therefore, we have to discuss how we can extend a spacetime to a family of Ehlers spacetimes. Only then we can talk about a Newtonian limit of this spacetime.

Definition 2.2.1 Let (M,g) be a spacetime. An *extension of the spacetime* (M,g) is a by $\lambda \in (0,1]$ parametrized family of Ehlers spacetimes

$$M_\lambda = (M, g(\lambda), h(\lambda), \nabla(\lambda), T(\lambda), \lambda)$$

with the additional condition

$$g(1) = -g.$$

Remark 2.2.2   a) As an extension is a family of Ehlers spacetimes this definition guarantees that it satisfies the conditions of the frame theory.

b) The additional condition

$$g(1) = -g$$

ensures that we can regain the original spacetime for $\lambda = 1$.

c) Observe that in definition (2.2.1) we talk about **an** extension of spacetime. We want to point out that we have not found a canonical way to extend a given spacetime.

2.3 Examples

In this section we want to discuss the Newtonian limits of some solutions of Einstein's field equations.

2.3.1 Minkowski spacetime

The Minkowski spacetime is the simplest solution of Einstein's field equations. We want to discuss one possibility of finding a Newtonian limit of Minkowski spacetime.

Definition 2.3.1   a) A *Minkowski space* is a vector space V of dimension $n \geq 2$ together with a non-degenerate, symmetric bilinear form g of index $n-1$. We call g a *Minkowski scalar product*.

b) A spacetime (M, g) is called *Minkowski spacetime* if (M, g) is isometrical to a four-dimensional Minkowski space.

We now consider the manifold $M = \mathbb{R}^4$ with a Minkowski scalar product g. In cartesian coordinates t, x^1, x^2 and x^3 it is of the following form:

$$(g_{ij}) = \text{diag}(-1, 1, 1, 1).$$

If one prefers the notation with the help of line elements this equals

$$ds^2 = -dt^2 + d(x^1)^2 + d(x^2)^2 + d(x^3)^2.$$

In order to introduce $\lambda \in (0, 1]$ so that the extended metrics satisfy the axioms of the frame theory we consider

$$\Phi_\lambda : \mathbb{R} \to \mathbb{R}, \quad t \mapsto \frac{1}{\sqrt{\lambda}} t,$$

which is a diffeomorphism from $\mathbb{R}$ to $\mathbb{R}$ for $\lambda \in (0, 1]$. Then the line element of $\tilde{g} = \Phi_\lambda^* g$ has the form

$$\widetilde{ds^2} = -\frac{1}{\lambda} dt^2 + d(x^1)^2 + d(x^2)^2 + d(x^3)^2.$$

This, of course, does not meet the conditions of the frame theory. For g we therefore multiply this line element with $-\lambda \neq 0$ for $\lambda \in (0, 1]$. Then we get

$$g : ds^2 = dt^2 - \lambda d(x^1)^2 - \lambda d(x^2)^2 - \lambda d(x^3)^2.$$

For h, we just have to invert $\widetilde{ds^2}$ and so we get

$$(g_{ij}(\lambda)) = \text{diag}(1, -\lambda, -\lambda, -\lambda) \text{ and } (h^{ij}(\lambda)) = \text{diag}(-\lambda, 1, 1, 1),$$

which satisfy the corresponding axioms of the frame theory in a natural way. For $\lambda \to 0$ we get the limits

$$(g_{ij}(0)) = \text{diag}(1, 0, 0, 0) \text{ and } (h^{ij}(0)) = \text{diag}(0, 1, 1, 1)$$

which also fulfil the conditions of the frame theory for $\lambda = 0$. The Christoffel symbols, of course, all equal zero and the Minkowski spacetime is flat. Thus the extended Minkowski spacetime has a genuine Newtonian limit with

$$\begin{aligned}
(g_{ij}(0)) &= \text{diag}(1, 0, 0, 0) \\
(h^{ij}(0)) &= \text{diag}(0, 1, 1, 1) \\
\Gamma^k_{ij}(0) &= 0, \quad \forall i, j, k \in \{0, 1, 2, 3\} \\
T^{ij} &= 0, \quad \forall i, j \in \{0, 1, 2, 3\}
\end{aligned}$$

and therefore

$$v = 0$$
$$f = 0$$
$$f = \mathrm{grad}(u)$$
$$\Rightarrow u = c, \quad \text{with } c = const.$$
$$\Delta u = 0$$
$$\Rightarrow \rho = 0.$$

Remark:

a) It is worth noting again that we have not found a canonical way to extend a solution of Einstein's field equations. Here we just describe one way of finding an extension which fulfils the conditions of the frame theory. Therefore, we do not know if the limit of any model of General Relativity is defined unambiguously. On the one hand we might find different extensions which have the same Newtonian limit. On the other hand it is not clear at all that we always get the same limit if we use a different extension.

b) Our Minkowski space $M = \mathbb{R}^4$ meets the topological conditions for a genuine Newtonian limit in a natural way. The manifold $M = \mathbb{R}^4$ is, of course, simply connected and geodesically complete in space direction.

2.3.2 The Schwarzschild solution

We first have a look at the Schwarzschild solution in its traditional form. In this case the Schwarzschild metric has the form

$$(g_{ij}) = \begin{pmatrix} -h(r) & 0 & 0 & 0 \\ 0 & h^{-1}(r) & 0 & 0 \\ 0 & 0 & r^2 & 0 \\ 0 & 0 & 0 & r^2 \sin^2 \vartheta \end{pmatrix}$$

with $h(r) = 1 - \frac{2m}{r}$, for $M = \mathbb{R} \times \mathbb{R}_+ \times_r \mathbb{S}^2$ with coordinates $t \in \mathbb{R}$, $r \in \mathbb{R}_+$ and the spherical coordinates ϑ and φ on $\mathbb{S}^2$.

We now use a similar way to introduce $\lambda \in (0, 1]$ as before. We again consider the diffeomorphism

$$\Phi_\lambda : \mathbb{R} \to \mathbb{R}, \quad t \mapsto \frac{1}{\sqrt{\lambda}} t.$$

Then we get from the line element of g,

$$ds^2 = -\left(1 - \frac{2m}{r}\right) dt^2 + \frac{1}{1 - \frac{2m}{r}} dr^2 + r^2 d\sigma^2,$$

the line element of $\widetilde{g} = \Phi_\lambda^* g$:

$$\widetilde{ds^2} = -\left(1 - \frac{2m}{r}\right) d\left(\frac{1}{\sqrt{\lambda}}t\right)^2 + \frac{1}{1 - \frac{2m}{r}} dr^2 + r^2 d\sigma^2$$

$$= -\frac{1}{\lambda}\left(1 - \frac{2m}{r}\right) dt^2 + \frac{1}{1 - \frac{2m}{r}} dr^2 + r^2 d\sigma^2.$$

If we now substitute m by λ, we get

$$ds^2 = -\frac{1}{\lambda}\left(1 - \frac{2\lambda}{r}\right) dt^2 + \frac{1}{1 - \frac{2\lambda}{r}} dr^2 + r^2 d\sigma^2,$$

with $0 < 2\lambda < r$. This means that

$$(\widetilde{g}_{ij}(\lambda)) = \text{diag}\left(-\frac{1}{\lambda}\left(1 - \frac{2\lambda}{r}\right), \frac{1}{1 - \frac{2\lambda}{r}}, r^2, r^2 \sin^2 \vartheta, \right).$$

Then we get the desired metrics by:

$$(g_{ij}(\lambda)) = -\lambda \cdot (\widetilde{g}_{ij}(\lambda))$$
$$(h^{ij}(\lambda)) = (\widetilde{g}_{ij}(\lambda))^{-1}.$$

As in the case of Minkowski spacetime,

$$(g_{ij}(\lambda)) = \text{diag}\left(1 - \frac{2\lambda}{r}, \frac{-\lambda}{1 - \frac{2\lambda}{r}}, -\lambda r^2, -\lambda r^2 \sin^2 \vartheta\right) \text{ and}$$

$$(h^{ij}(\lambda)) = \text{diag}\left(-\lambda\left(\frac{1}{1 - \frac{2\lambda}{r}}\right), 1 - \frac{2\lambda}{r}, \frac{1}{r^2}, \frac{1}{r^2 \sin^2 \vartheta}\right)$$

also meet the corresponding conditions of the frame theory. For $\lambda \in (0, 1]$ and $0 < 2\lambda < r$, g has rank 4 and index 3, h has rank 4 and index 1. For $\lambda \to 0$ we get the limits

$$(g_{ij}(0)) = \text{diag}\,(1, 0, 0, 0) \text{ and}$$
$$(h^{ij}(0)) = \text{diag}\left(0, 1, \frac{1}{r^2}, \frac{1}{r^2 \sin^2 \vartheta}\right).$$

Now we want to discuss the limit more closely. We first study the Christoffel symbols in case of $\lambda \in (0, 1]$, but also for the limit connection. We first have to note that $(g_{ij}(\lambda))$ is no Lorentzian metric, but $(h^{ij}(\lambda))$ has Lorentzian signature. But if we now want to calculate the Christoffel symbols for $\lambda > 0$ it does not matter if we use g or h as they are "quasi" inverse.

In Appendix B we calculate the Christoffel symbols, the components of the Ricci tensor and the divergence of the function f in detail. Here we just have a look at the results.

2.3. Examples

The Christoffel symbols of $(g_{ij}(\lambda))$ which do not vanish have the form:

$$\Gamma^1_{11} = \frac{\lambda}{2\lambda r - r^2}$$
$$\Gamma^1_{00} = \frac{r - 2\lambda}{r^3}$$
$$\Gamma^1_{22} = 2\lambda - r$$
$$\Gamma^1_{33} = (2\lambda - r)\sin^2 \vartheta$$
$$\Gamma^2_{33} = -\sin \vartheta \cos \vartheta$$
$$\Gamma^0_{10} = \frac{\lambda}{r^2 - 2\lambda r}$$
$$\Gamma^2_{12} = \frac{1}{r}$$
$$\Gamma^3_{13} = \frac{1}{r}$$
$$\Gamma^3_{23} = \frac{\cos \vartheta}{\sin \vartheta}.$$

Then the Christoffel symbols of the limit connection have the form:

$$\Gamma^1_{00} = \frac{1}{r^2}$$
$$\Gamma^1_{22} = -r$$
$$\Gamma^1_{33} = -r \sin^2 \vartheta$$
$$\Gamma^2_{33} = -\sin \vartheta \cos \vartheta$$
$$\Gamma^2_{12} = \frac{1}{r}$$
$$\Gamma^3_{13} = \frac{1}{r}$$
$$\Gamma^3_{23} = \frac{\cos \vartheta}{\sin \vartheta}$$
$$\Gamma^k_{ij} = 0, \text{ otherwise.}$$

There is no component of the Ricci tensor that does not vanish. Therefore,

$$\text{Ric} = 0.$$

Then our extended Schwarzschild spacetime has a genuine Newtonian limit with

$$(g_{ij}(0)) = \text{diag}(1, 0, 0, 0)$$
$$(h^{ij}(0)) = \text{diag}\left(0, 1, \frac{1}{r^2}, \frac{1}{r^2 \sin^2 \vartheta}\right)$$
$$\Gamma^k_{ij}(0) \text{ as above}$$
$$T^{ij} = 0, \quad \forall i, j \in \{0, 1, 2, 3\}$$

and so

$$v = 0$$
$$f = \left(\frac{1}{r^2}, 0, 0\right)$$
$$f = \text{grad}(u)$$
$$\Rightarrow u = -\frac{1}{r} + c, \text{ with } c = \text{const.}$$
$$\rho = 0$$
$$\Delta u = 0.$$

Remark: For the extended Schwarzschild solution the manifold $M = \mathbb{R} \times \mathbb{R}_+ \times \mathbb{S}^2$ is simply connected as all components are simply connected. But it is not geodesically complete in space direction as we only consider a solution outside the star which is considered to be in the origin of the spacetime. In order to achieve geodesic completeness we also have to consider the inner Schwarzschild solution and connect the two solutions.

2.3.3 The Kerr Solution

Let us consider the Kerr Solution in its original form. In Boyer-Lindquist coordinates $t \in \mathbb{R}$, $r \in \mathbb{R}_+$, $\vartheta \in (0, \pi)$ and $\varphi \in \mathbb{S}^1$ we have:

$$(g_{ij}) = \begin{pmatrix} -1 + \frac{2mr}{\rho} & 0 & 0 & -\frac{2mra\sin^2\vartheta}{\rho} \\ 0 & \frac{\rho}{\Delta} & 0 & 0 \\ 0 & 0 & \rho & 0 \\ -\frac{2mra\sin^2\vartheta}{\rho} & 0 & 0 & \left(r^2 + a^2 + \frac{2mra^2\sin^2\vartheta}{\rho}\right)\sin^2\vartheta \end{pmatrix}$$

with $\rho := r^2 + a^2\cos^2\vartheta$ and $\Delta := r^2 - 2mr + a^2$. Usually the two constants m and a are called mass and angular momentum respectively. Now we are going to copy the procedure of the Schwarzschild case. We first introduce $\lambda \neq 0$ with the help of the diffeomorphism

$$\Phi_\lambda : \mathbb{R} \to \mathbb{R}, \quad t \mapsto \frac{1}{\sqrt{\lambda}} t$$

and we get for $\tilde{g} = \Phi_\lambda^* g$

$$(\tilde{g}_{ij}) = \begin{pmatrix} -\frac{1}{\lambda}\left(1 - \frac{2mr}{\rho}\right) & 0 & 0 & \frac{-2mra\sin^2\vartheta}{\sqrt{\lambda}\rho} \\ 0 & \frac{\rho}{\Delta} & 0 & 0 \\ 0 & 0 & \rho & 0 \\ \frac{-2mra\sin^2\vartheta}{\sqrt{\lambda}\rho} & 0 & 0 & \left(r^2 + a^2 + \frac{2mra^2\sin^2\vartheta}{\rho}\right)\sin^2\vartheta \end{pmatrix}.$$

2.3. Examples

We then get $(g_{ij}(\lambda))$ by multiplying $(\tilde{g}_{ij})$ with $(-\lambda) \neq 0$ and $(h_{ij}(\lambda))$ by inverting $(\tilde{g}_{ij})$. Furthermore, we replace m by λ. This means:

$$(g_{ij}(\lambda)) = \begin{pmatrix} 1 - \frac{2\lambda r}{\rho} & 0 & 0 & 2\lambda^{\frac{3}{2}}\frac{ra\sin^2\vartheta}{\rho} \\ 0 & -\lambda\frac{\rho}{\tilde{\Delta}} & 0 & 0 \\ 0 & 0 & -\lambda\rho & 0 \\ 2\lambda^{\frac{3}{2}}\frac{ra\sin^2\vartheta}{\rho} & 0 & 0 & -\lambda\left(r^2 + a^2 + \frac{2\lambda ra^2\sin^2\vartheta}{\rho}\right)\sin^2\vartheta \end{pmatrix},$$

with $\rho := r^2 + a^2 \cos^2 \vartheta$ and $\tilde{\Delta} := r^2 - 2\lambda r + a^2$.

The inversion of $(\tilde{g}_{ij})$ is carried out in detail in Appendix B. Here we have the result:

$$(h^{ij}(\lambda)) = \begin{pmatrix} -\lambda\left(1 + \frac{2mr(r^2+a^2)}{\tilde{\Delta}\rho}\right) & 0 & 0 & -\sqrt{\lambda}\frac{2mra}{\rho\tilde{\Delta}} \\ 0 & \frac{\tilde{\Delta}}{\rho} & 0 & 0 \\ 0 & 0 & \frac{1}{\rho} & 0 \\ -\sqrt{\lambda}\frac{2mra}{\rho\tilde{\Delta}} & 0 & 0 & \frac{\rho - 2mr}{\sin^2\vartheta \cdot \rho\tilde{\Delta}} \end{pmatrix}$$

with $\rho := r^2 + a^2 \cos^2 \vartheta$ and $\tilde{\Delta} := r^2 - 2\lambda r + a^2$. As in the cases before $(g_{ij}(\lambda))$ and $(h^{ij}(\lambda))$ meet the conditions of the frame theory: for $\lambda \in (0,1]$, g has rank 4 and index 3, h has rank 4 and index 1. For $\lambda \to 0$ we get the following limits:

$$(g_{ij}(0)) = \text{diag}(1, 0, 0, 0),$$
$$(h^{ij}(0)) = \text{diag}\left(0, \frac{r^2 + a^2}{r^2 + a^2 \cos^2 \vartheta}, \frac{1}{r^2 + a^2 \cos^2 \vartheta}, \frac{1}{(r^2 + a^2)\sin^2 \vartheta}\right).$$

Now we again discuss the limit more closely. We first want to study the Christoffel symbols in case of $\lambda \in (0, 1]$, but also for the limit connection. As before you can look at the exact calculations in Appendix B. For the divergence of f (in order to show that $\Delta u = 0$) one has to copy the calculations of the Schwarzschild case, which is not carried out in detail.

The Christoffel symbols of $(g_{ij}(\lambda))$ which do not vanish have the form:

$$\Gamma^1_{00} = \frac{\tilde{\Delta}(r^2 - a^2 \cos^2 \vartheta)}{\rho^3}$$

$$\Gamma^2_{00} = \frac{-2ra^2 \sin\vartheta \cos\vartheta}{\rho^3}$$

$$\Gamma^0_{01} = \Gamma^0_{10} = \frac{\lambda(r^2 - a^2 \cos^2 \vartheta)(r^2 + a^2)}{\rho^2 \tilde{\Delta}}$$

$$\Gamma^0_{02} = \Gamma^0_{20} = \frac{-4\lambda ra^2 \sin\vartheta \cos\vartheta}{\rho^2}$$

$$\Gamma^3_{01} = \Gamma^3_{10} = \frac{\sqrt{\lambda}a(r^2 - a^2 \cos^2 \vartheta)}{\tilde{\Delta}\rho^2}$$

$$\Gamma^3_{02} = \Gamma^3_{20} = \frac{-2\sqrt{\lambda}ra\cos\vartheta}{\sin\vartheta \cdot \rho^2}$$

$$\Gamma^0_{13} = \Gamma^0_{31} = \frac{-\lambda^{\frac{3}{2}}a\sin^2\vartheta}{\rho^2\widetilde{\Delta}}[(r^2+a^2)\cdot(r^2-a^2\cos^2\vartheta)+2r^2\rho]$$

$$\Gamma^0_{23} = \Gamma^0_{32} = \frac{2\lambda^{\frac{3}{2}}ra^3\sin^3\vartheta\cos\vartheta}{\rho^2}$$

$$\Gamma^1_{30} = \Gamma^1_{03} = -\frac{\sqrt{\lambda\widetilde{\Delta}}\cdot a\sin^2\vartheta(r^2-a^2\cos^2\vartheta)}{\rho^3}$$

$$\Gamma^2_{30} = \Gamma^2_{03} = \frac{2\sqrt{\lambda}ra\sin\vartheta\cos\vartheta(r^2+a^2)}{\rho^3}$$

$$\Gamma^1_{11} = \frac{ra^2\sin^2\vartheta - \lambda(r^2-a^2\cos^2\vartheta)}{\rho\widetilde{\Delta}}$$

$$\Gamma^2_{11} = \frac{a^2\sin\vartheta\cos\vartheta}{\rho\widetilde{\Delta}}$$

$$\Gamma^1_{21} = \Gamma^1_{12} = \frac{-a^2\sin\vartheta\cos\vartheta}{\rho}$$

$$\Gamma^1_{22} = -\frac{\widetilde{\Delta}r}{\rho}$$

$$\Gamma^2_{21} = \Gamma^2_{12} = \frac{r}{\rho}$$

$$\Gamma^2_{22} = -\frac{a^2\cos\vartheta\sin\vartheta}{\rho}$$

$$\Gamma^1_{33} = -\frac{\widetilde{\Delta}\sin^2\vartheta(r\rho^2 - \lambda a^2\sin^2\vartheta(r^2-a^2\cos\vartheta))}{\rho^3}$$

$$\Gamma^2_{33} = -\frac{\sin\vartheta\cos\vartheta(\widetilde{\Delta}\rho^2 + 2\lambda r(r^2+a^2)^2)}{\rho^3}$$

$$\Gamma^3_{31} = \Gamma^3_{13} = \frac{-\lambda(2r^2\rho - a^2\sin^2\vartheta(r^2-a^2\cos\vartheta)) + r\rho^2}{\widetilde{\Delta}\rho^2}$$

$$\Gamma^3_{32} = \Gamma^3_{23} = \frac{\cos\vartheta(2\lambda ra^2\sin^2\vartheta + \rho^2)}{\sin\vartheta \cdot \rho^2}.$$

Then the Christoffel symbols of the limit connection have the form:

$$\Gamma^1_{00} = \frac{(r^2+a^2)(r^2-a^2\cos^2\vartheta)}{(r^2+a^2\cos^2\vartheta)^3}$$

$$\Gamma^2_{00} = \frac{-2ra^2\sin\vartheta\cos\vartheta}{(r^2+a^2\cos^2\vartheta)^3}$$

$$\Gamma^1_{11} = \frac{ra^2\sin^2\vartheta}{(r^2+a^2\cos^2\vartheta)(r^2+a^2)}$$

$$\Gamma^2_{11} = \frac{a^2\sin\vartheta\cos\vartheta}{(r^2+a^2\cos^2\vartheta)(r^2+a^2)}$$

$$\Gamma^1_{21} = \Gamma^1_{12} = \frac{-a^2\sin\vartheta\cos\vartheta}{r^2+a^2\cos^2\vartheta}$$

$$\Gamma^1_{22} = -\frac{(a^2+r^2)r}{r^2+a^2\cos^2\vartheta}$$

$$\Gamma^2_{21} = \Gamma^2_{12} = \frac{r}{r^2+a^2\cos^2\vartheta}$$

$$\Gamma^2_{22} = -\frac{a^2\cos\vartheta\sin\vartheta}{r^2+a^2\cos^2\vartheta}$$

$$\Gamma^1_{33} = -\frac{r\sin^2\vartheta(r^2+a^2)}{r^2+a^2\cos^2\vartheta}$$

$$\Gamma^2_{33} = -\frac{\sin\vartheta\cos\vartheta(r^2+a^2)}{r^2+a^2\cos^2\vartheta}$$

$$\Gamma^3_{31} = \Gamma^3_{13} = \frac{r}{r^2+a^2}$$

$$\Gamma^3_{32} = \Gamma^3_{23} = \frac{\cos\vartheta}{\sin\vartheta}$$

$$\Gamma^k_{ij} = 0, \text{ otherwise.}$$

There is no component of the Ricci tensor that does not vanish. Therefore,

$$\text{Ric} = 0.$$

Then our extended Kerr spacetime has a genuine Newtonian limit with

$$(g_{ij}(0)) = \text{diag}(1,0,0,0),$$

$$(h^{ij}(0)) = \text{diag}\left(0, \frac{r^2+a^2}{r^2+a^2\cos^2\vartheta}, \frac{1}{r^2+a^2\cos^2\vartheta}, \frac{1}{(r^2+a^2)\sin^2\vartheta}\right),$$

$$\Gamma^k_{ij}(0) \text{ as above}$$

$$T^{ij} = 0, \quad \forall i,j \in \{0,1,2,3\}$$

and so

$$v = 0$$
$$f = (\Gamma^1_{00}, \Gamma^2_{00}, 0)$$
$$f = \text{grad}(u)$$
$$\rho = 0$$
$$\Delta u = 0.$$

Remark 2.3.2 Of course, our standard examples are extensions in the above sense. We already know that the corresponding families of Ehlers spacetimes satisfy the conditions of the frame theory. Therefore, we just have to consider the additional condition. Here we assume that the mass m equals one, $m = 1$:

a) Minkowski spacetime: Originally the Minkowski metric has the form

$$(g_{ij}) = \text{diag}(-1,1,1,1).$$

After extension we had
$$(g_{ij}(\lambda)) = \mathrm{diag}(1, -\lambda, -\lambda, -\lambda),$$
which means that the condition
$$g = -g(1)$$
is satisfied as we can immediately see.

b) **Schwarzschild spacetime:** The Schwarzschild metric is of the form
$$(g_{ij}) = \mathrm{diag}(-h(r), h^{-1}(r), r^2, r^2 \sin^2 \vartheta),$$
with $h(r) = 1 - \frac{2m}{r}$. After extension we get
$$(g_{ij}(\lambda)) = \mathrm{diag}\left(1 - \frac{2\lambda}{r}, \frac{-\lambda}{1 - \frac{2\lambda}{r}}, -\lambda r^2, -\lambda r^2 \sin^2 \vartheta\right)$$
and therefore for $\lambda = 1$:
$$(g_{ij}(1)) = \mathrm{diag}\left(1 - \frac{2}{r}, \frac{-1}{1 - \frac{2}{r}}, -r^2, -r^2 \sin^2 \vartheta\right),$$
which equals $-g$ for $m = 1$.

c) **Kerr spacetime:** The Kerr metric is of the form
$$(g_{ij}) = \begin{pmatrix} -1 + \frac{2mr}{\rho} & 0 & 0 & -\frac{2mra \sin^2 \vartheta}{\rho} \\ 0 & \frac{\rho}{\Delta} & 0 & 0 \\ 0 & 0 & \rho & 0 \\ -\frac{2mra \sin^2 \vartheta}{\rho} & 0 & 0 & (r^2 + a^2 + \frac{2mra^2 \sin^2 \vartheta}{\rho}) \sin^2 \vartheta \end{pmatrix}$$
with $\rho := r^2 + a^2 \cos^2 \vartheta$ and $\Delta := r^2 - 2mr + a^2$. The extension mentioned above has the form
$$(g_{ij}(\lambda)) = \begin{pmatrix} 1 - \frac{2\lambda r}{\rho} & 0 & 0 & 2\lambda^{\frac{3}{2}} \frac{ra \sin^2 \vartheta}{\rho} \\ 0 & -\lambda \frac{\rho}{\tilde{\Delta}} & 0 & 0 \\ 0 & 0 & -\lambda \rho & 0 \\ 2\lambda^{\frac{3}{2}} \frac{ra \sin^2 \vartheta}{\rho} & 0 & 0 & -\lambda \left(r^2 + a^2 + \frac{2\lambda ra^2 \sin^2 \vartheta}{\rho}\right) \sin^2 \vartheta \end{pmatrix}$$
with $\rho := r^2 + a^2 \cos^2 \vartheta$ and $\tilde{\Delta} := r^2 - 2\lambda r + a^2$. This means that we get for $\lambda = 1$:
$$(g_{ij}(1)) = \begin{pmatrix} 1 - \frac{2r}{\rho} & 0 & 0 & \frac{2ra \sin^2 \vartheta}{\rho} \\ 0 & -\frac{\rho}{\Delta} & 0 & 0 \\ 0 & 0 & -\rho & 0 \\ \frac{2ra \sin^2 \vartheta}{\rho} & 0 & 0 & -\left(r^2 + a^2 + \frac{2ra^2 \sin^2 \vartheta}{\rho}\right) \sin^2 \vartheta \end{pmatrix}$$

with $\rho := r^2 + a^2 \cos^2 \vartheta$ and $\overline{\Delta} := r^2 - 2r + a^2$. This again means that

$$g(1) = -g$$

for $m = 1$.

2.4 Existence of a limit

In this section we want to discuss conditions which ensure the existence of a limit of a family of Ehlers spacetimes. Some conditions are already known, see, for example [31], §3.2, but we also find a more geometrical equivalent of the conditions mentioned in [31].

2.4.1 The limit of the metrics

We start with the limit of the metrics $g(\lambda)$ and $h(\lambda)$. Therefore, we first discuss the geometry of the vector space of symmetric bilinear forms. Although we have a special situation ($V = TM_p$, $\dim(V) = 4$), we carry out a general discussion of the vector space of symmetric bilinear forms: Let V be a finite dimensional real vector space, $n = \dim(V)$ and $W = \operatorname{Sym}_2 V$ the vector space of symmetric bilinear forms on V. We consider the *degenerate cone*

$$C := \{s \in W : \operatorname{Deg}(s) = \{v \in V : s(v, w) = 0, \forall w \in V\} \neq 0\}.$$

Note, that the name "cone" is derived from the fact that for $s \in C$ also $\lambda s \in C$, for all $\lambda \in \mathbb{R}$. We now define the subset $C^+ \subseteq C$ by

$$C^+ := \{s \in W : s \geq 0,\ \dim(\operatorname{Deg}(s)) = 1\} \subseteq C.$$

Proposition 2.4.1 *$C^+ \subseteq W$ is a hypersurface, which means that an open neighbourhood $U \subseteq W$ of s_0 and a map $f : U \to \mathbb{R}$ exist, for all $s_0 \in C^+$, so that*

(i) $C^+ \cap U = \{s \in U : f(s) = 0\}$;

(ii) $df(s_0) \neq 0$.

Proof. Let $s_0 \in C^+$. According to the theorem of Sylvester (see (0.3.6)) there is a basis $\mathfrak{A}$ of V so that

$$M(s_0; \mathfrak{A}) = \operatorname{diag}(0, 1, ..., 1).$$

Without restriction let therefore $W = \operatorname{Sym}_2(\mathbb{R}^n)$ and $s_0 = \operatorname{diag}(0, 1, ..., 1)$. Furthermore, we define

$$f : \operatorname{Sym}_2(\mathbb{R}^n) \to \mathbb{R},\ s \mapsto \det(s).$$

If now $s \in W$ and

$$s = \begin{pmatrix} * & * \\ * & A^{11} \end{pmatrix},$$

$A^{11} \in \text{Mat}_{n-1}(\mathbb{R})$, then there is an open neighbourhood $U \subseteq W$ of s_0 so that $\det(A^{11}) \neq 0$, for all $s \in U$. For $s \in f^{-1}(0) \subseteq U$ we therefore have $\dim(\text{Deg}(s)) = 1$ and $s \geq 0$, thus $s \in U \cap C^+$ and $f^{-1}(0) \subseteq U \cap C^+$.

But as $C^+ \subseteq C$, we have $U \cap C^+ \subseteq f^{-1}(0)$, of course, and thus

$$U \cap C^+ = \{s \in U : f(s) = 0\}.$$

Finally,

$$\frac{\partial f}{\partial x_{11}}(s_0) = \left.\frac{d}{dt}\right|_{t=0} (\det(\text{diag}(t, 1, ..., 1))) = 1 \neq 0$$

holds and therefore $df(s_0) \neq 0$. □

Theorem 2.4.2 *Let $s \in C^+$ and $v \in \text{Deg}(s) \setminus \{0\}$. Then the following holds:*

$$TC_s^+ = \{w \in W : w(v, v) = 0\}.$$

Proof. Let $s_0 \in C^+$ be arbitrary. Again, we use that $V = \mathbb{R}^n$ and, without restriction, $s_0 = \text{diag}(0, 1, ..., 1)$. As above, we therefore get

$$\text{grad}(f)(s_0) = \text{diag}(1, 0, ..., 0),$$

since:

$$\frac{\partial f}{\partial x_{11}}(s_0) = \left.\frac{d}{dt}\right|_{t=0} (\det(\text{diag}(t, 1, ..., 1))) = 1$$

$$\frac{\partial f}{\partial x_{ii}}(s_0) = \left.\frac{d}{dt}\right|_{t=0} (\det(\text{diag}(0, 1, ..., 1, t, 1, ..., 1))) = 0$$

$$\frac{\partial f}{\partial x_{jk}}(s_0) = \left.\frac{d}{dt}\right|_{t=0} (\det(\text{diag}(0, 1, ..., 1) + t \cdot E_{jk})) = 0,$$

for $i = 2, ..., n$ and $1 \leq j, k \leq n$ and $j \neq k$. Here E_{jk} is the matrix whose components are zero, except for the component e_{jk} which equals one, $e_{jk} = 1$. Thus we have

$$TC_{s_0}^+ = \ker(\text{grad}(f)(s_0)) = \{w = (w_{ij}) \in W : w_{11} = 0\}.$$

But, of course,

$$w_{11} = w(e_1, e_1)$$

and $\text{Deg}(s_0) = \mathbb{R}e_1$. This shows the claim. □

After this basic discussion of the vector space of the symmetric bilinear forms, we now start discussing the existence of the limit. Therefore, let V be a real, finite dimensional vector space, $\dim V = n$ and $W = \text{Sym}_2 V$. Let $h : [0, 1] \to W$ be a smooth curve so that $h(\lambda) \notin C$, for all $\lambda \in (0, 1]$ and $h(0) \in C^+$. Let $\psi : [0, 1] \to \text{Hom}(V, V^*)$ be given by $\lambda \mapsto \psi_\lambda$ with $\langle \psi_\lambda v, w \rangle := h_\lambda(v, w)$. Then ψ_λ is invertible for $\lambda \in (0, 1]$ and we can consider the dual curve

$g : (0,1] \to \mathrm{Sym}_2(V^*)$, given by

$$g_\lambda(\alpha, \beta) := -\lambda h_\lambda(\psi_\lambda^{-1}\alpha, \psi_\lambda^{-1}\beta),$$

for $\alpha, \beta \in V^*$. The following holds:

Proposition 2.4.3 *Suppose that $g_0 := \lim_{\lambda \to 0} g_\lambda$ exists and $g : [0,1] \to \mathrm{Sym}_2(V^*)$ is differentiable. Then the following holds:*

a) $g_0 \neq 0$ and $h'(0) \notin TC^+_{h(0)}$;

b) $\mathrm{Deg}\,(g_0) = \mathrm{Ann}\,(\mathrm{Deg}\,(h_0))$;

c) *If $\mathrm{rk}\,(h_\lambda) = n$ and $\mathrm{index}\,(h_\lambda) = 1$ for $\lambda \in (0,1]$, then $\mathrm{rk}\,(g_0) = 1$ and $\mathrm{index}\,(g_0) = 0$.*

Proof. First we define $\varphi : [0,1] \to \mathrm{Hom}\,(V^*, V)$ by

$$\langle \alpha, \varphi_\lambda \beta \rangle := g_\lambda(\alpha, \beta)$$

for $\alpha, \beta \in V^*$. Then φ is continuous and for $\lambda \in (0,1]$ the following holds:

$$\langle \alpha, \varphi_\lambda \circ \psi_\lambda(v) \rangle = g_\lambda(\alpha, \psi_\lambda v) = -\lambda h_\lambda(\psi_\lambda^{-1}\alpha, v) = -\lambda \langle \alpha, v \rangle = \langle \alpha, (-\lambda \,\mathrm{id}_V)(v) \rangle,$$

for all $\alpha \in V^*$ and for all $v \in V$. Thus

$$\varphi_\lambda \circ \psi_\lambda = -\lambda \,\mathrm{id}_V. \qquad (2.1)$$

As $\lambda \in (0,1]$ it also holds:

$$\psi_\lambda \circ \varphi_\lambda = -\lambda \,\mathrm{id}_{V^*}. \qquad (2.2)$$

If we consider the calculations above, we see that for the limit $\lambda \to 0$, (2.1) also holds and results in

$$\varphi_0 \circ \psi_0 = 0. \qquad (2.3)$$

Now we want to show that

$$\mathrm{im}(\psi_0) = \mathrm{Ann}(\mathrm{Deg}(h_0)).$$

We first observe that

$$\langle \psi_0(v), w \rangle = h_0(v, w) = 0,$$

for all $w \in \mathrm{Deg}(h_0)$ and $v \in V$ arbitrary, which means that

$$\mathrm{im}(\psi_0) \subseteq \mathrm{Ann}(\mathrm{Deg}(h_0)).$$

But both have the same dimension, since

$$\dim(\mathrm{Ann}(\mathrm{Deg}(h_0))) = n - \dim(\mathrm{Deg}(h_0)) = n - \dim(\ker(\psi_0)) = \dim(\mathrm{im}(\psi_0)),$$

because of $\ker(\psi_0) = \operatorname{Deg}(h_0)$. From (2.3) now follows:

$$\operatorname{Ann}(\operatorname{Deg}(h_0)) = \operatorname{im}(\psi_0) \subseteq \ker(\varphi_0) = \operatorname{Deg}(g_0).$$

As $\dim(\operatorname{Deg}(h_0)) = 1$, the equality holds if we show that $g_0 \neq 0$. But from

$$\varphi(\lambda)\psi(\lambda) = -\lambda \operatorname{id}_V,$$

for all $\lambda \in [0,1]$, follows after differentiation and for $\lambda = 0$:

$$-\operatorname{id}_V = \varphi'(0)\psi_0 + \varphi_0\psi'(0). \tag{2.4}$$

If we suppose that $g_0 = 0$, then $\varphi_0 = 0$ and therefore ψ_0 is invertible, but it is not invertible, of course. Thus $g_0 \neq 0$ and therefore

$$\operatorname{Ann}(\operatorname{Deg}(h_0)) = \operatorname{Deg}(g_0),$$

which shows (b).

Now we can show that from (2.3)

$$\psi_0 \circ \varphi_0 = 0 \tag{2.5}$$

also follows (remember (1.2.2), where we showed this for $n = 4$, but the proof is the same). But this means that $\operatorname{im}(\varphi_0) \subseteq \operatorname{Deg}(h_0)$. But we even have $\operatorname{im}(\varphi_0) = \operatorname{Deg}(h_0)$, since

$$\ker(\varphi_0) = \operatorname{Deg}(g_0) = \operatorname{Ann}(\operatorname{Deg}(h_0)) = \operatorname{im}(\psi_0)$$

and so

$$\operatorname{Deg}(h_0) = \ker(\psi_0) = n - \operatorname{im}(\psi_0) = n - \ker(\varphi_0) = \operatorname{im}(\varphi_0).$$

Thus, let $v_0 \in \operatorname{Deg}(h_0)\setminus\{0\}$. Then there is an $\alpha_0 \in V^*$ with $v_0 = \varphi_0(\alpha_0)$ and the following holds:

$$h'_0(v_0, v_0) = \langle \psi'_0 v_0, v_0 \rangle = \langle \psi'_0 v_0, \varphi_0(\alpha_0) \rangle = g_0(\psi'_0 v_0, \alpha_0) = g_0(\alpha_0, \psi'_0 v_0) = \langle \alpha_0, \varphi_0 \circ \psi'_0(v_0) \rangle$$
$$\stackrel{(2.4)}{=} \langle \alpha_0, -v_0 \rangle - \underbrace{\langle \alpha_0, \varphi'_0 \psi_0 v_0 \rangle}_{=0} = -\langle \alpha_0, \varphi(\alpha_0) \rangle = -g_0(\alpha_0, \alpha_0).$$

Since $g_0(\alpha_0, \alpha_0) \neq 0$, (remember that if $g_0(\alpha_0, \alpha_0) = 0$, then $g_0 = 0$, but $g_0 \neq 0$), it follows (with the help of (2.4.2)) that

$$h'_0 \notin TC^+_{h_0}.$$

This means that (a) holds.

Let finally $\operatorname{rk}(h_\lambda) = n$ and $\operatorname{index}(h_\lambda) = 1$. Then $h_\lambda(v_0, v_0) < 0$ for $0 < \lambda < \epsilon$ and ϵ small enough (otherwise h_λ would be positive definite here). But then

$$h'_0(v_0, v_0) = \lim_{\lambda \to 0} \underbrace{\frac{1}{\lambda}}_{>0} \underbrace{h_\lambda(v_0, v_0)}_{<0} \leq 0$$

holds, which means that in this case $h'_0(v_0, v_0) < 0$. Thus, $g_0(\alpha_0, \alpha_0) > 0$ and therefore g_0 has rank 1 and index 0. But this shows (c) and the claim. $\square$

Lemma 2.4.4 *Let V be a real finite dimensional vector space, $\dim V = n$ and $W = Sym_2 V$. Let $h : [0, 1] \to W$ be a smooth curve so that $h(\lambda) \notin C$, for all $\lambda \in (0, 1]$ and $h(0) \in C^+$. Then $h'_0 \notin TC^+_{h_0}$ if and only if the following holds: for $A(\lambda) = M(h(\lambda); \mathfrak{A})$,*

$$\left.\frac{d}{d\lambda}\right|_{\lambda=0} \det(A(\lambda)) \neq 0$$

holds. Here $\mathfrak{A}$ is an arbitrary basis of V (and the claim does not depend on the choice of the basis).

Proof. Again we suppose that $V = \mathbb{R}^n$ and $h(0) = \text{diag}(0, 1, ..., 1) = A(0)$, without restriction. Then we have, due to the Leibniz formula,

$$\left.\frac{d}{d\lambda}\right|_{\lambda=0} \det(A(\lambda)) = \left.\frac{d}{d\lambda}\right|_{\lambda=0} \left(\sum_{j=1}^n (-1)^{1+j} a_{1j}(\lambda) \det(A^{1j}(\lambda))\right)$$

$$= \sum_{j=1}^n (-1)^{j+1} (a'_{1j}(0)) \underbrace{\det(A^{1j})(0)}_{=0 \text{ for } j \geq 2} + \underbrace{a_{1j}(0)}_{=0 \text{ for all } j} \det(A^{1j})'(0))$$

$$= a'_{11}(0) \det(E_{n-1}) = a'_{11}(0) = h'_0(e_1, e_1).$$

According to (2.4.2) we know that $h'_0(e_1, e_1) \neq 0$ if and only if $h'_0 \notin TC^+_{h_0}$ and this shows the claim. $\square$

Theorem 2.4.5 *Let V be a real finite dimensional vector space, $\dim V = n$ and $W = Sym_2 V$. Let $h : [0, 1] \to W$ be a smooth curve so that $h(\lambda) \notin C$, for all $\lambda \in (0, 1]$ and $h(0) \in C^+$. Let $\psi : [0, 1] \to Hom(V, V^*)$ be given by $\lambda \mapsto \psi_\lambda$ with*

$$\langle \psi_\lambda v, w \rangle := h_\lambda(v, w).$$

Again we can consider the dual curve $g : (0, 1] \to Sym_2(V^*)$, given by

$$g_\lambda(\alpha, \beta) := -\lambda h_\lambda(\psi_\lambda^{-1}\alpha, \psi_\lambda^{-1}\beta),$$

for $\alpha, \beta \in V^*$.
If $h'_0 \notin TC^+_{h_0}$, then there is a limit $g(0) := \lim_{\lambda \to 0} g(\lambda)$ and the continued curve $g : [0, 1] \to Sym_2(V^*)$ is smooth if $h : [0, 1] \to Sym_2(V)$ is smooth.

2.4. Existence of a limit

Proof. Let $\mathfrak{A}$ be a basis of V, without restriction, $A(0) = \mathrm{diag}(0, 1, ..., 1)$ and

$$A(\lambda) = M(h(\lambda); \mathfrak{A}), \quad B(\lambda) = M(g(\lambda); \mathfrak{A}^*).$$

But then we also have

$$M(\psi(\lambda); \mathfrak{A}, \mathfrak{A}^*) = A(\lambda), \quad M(\varphi(\lambda); \mathfrak{A}^*, \mathfrak{A}) = B(\lambda)$$

and $g(\lambda)$ converges if and only if $B(\lambda)$ converges. Now from $\varphi_\lambda \circ \psi_\lambda = -\lambda \, \mathrm{id}_V$ it follows that $A(\lambda) B(\lambda) = -\lambda E_n$, thus

$$B(\lambda) = -\lambda A^{-1}(\lambda) = -\frac{\lambda}{\det(A(\lambda))} \cdot \mathrm{adj}(A(\lambda))$$

for $\lambda \neq 0$. If now $h_0' \notin TC_{h_0}^+$ it follows that

$$\lim_{\lambda \to 0} \frac{\det(A(\lambda))}{\lambda} = \frac{d}{d\lambda}\bigg|_{\lambda=0} \det(A(\lambda))$$

exists and is not zero due to lemma (2.4.4). But this means that $\lim_{\lambda \to 0} B(\lambda)$ also exists (since, of course, $\mathrm{adj}(A(0))$ exists, as $A(0)$ exists and the components of $\mathrm{adj}(A(0))$ are only sums and products of the components of $A(0)$) and we have:

$$B_0 := \lim_{\lambda \to 0} B(\lambda) = \frac{-1}{\det(A)'(0)} \mathrm{adj}(A(0)) = \frac{-1}{\det(A)'(0)} \mathrm{diag}(1, 0, ..., 0).$$

Finally, we want to show that $g : [0, 1] \to \mathrm{Sym}_2(V^*)$ is smooth. We therefore consider the following developments (remember that $h : [0, 1] \to W$ is smooth):

$$A(\lambda) = A(0) + \lambda \cdot A'(0) + o(\lambda)$$
$$\mathrm{adj}(A(\lambda)) = \mathrm{adj}(A(0)) + \mathrm{adj}(A)'(0) \cdot \lambda + o(\lambda)$$
$$\det(A(\lambda)) = \underbrace{\det(A(0))}_{=0} + \underbrace{\det(A)'(0)}_{\neq 0} \cdot \lambda + o(\lambda)$$
$$\frac{\det(A(\lambda))}{\lambda} = \det(A)'(0) + o(1) \Rightarrow \frac{\lambda}{\det(A(\lambda))} = \frac{1}{\det(A)'(0)} + o(1).$$

Therefore, the following holds:

$$\frac{1}{\lambda}(B(\lambda) - B(0)) = \frac{1}{\lambda}\left(-\frac{\lambda}{\det(A(\lambda))} \cdot \mathrm{adj}(A(\lambda)) + \frac{1}{\det(A)'(0)} \cdot \mathrm{adj}(A(0))\right)$$
$$= \frac{1}{\lambda}\left(-\left(\frac{1}{\det(A)'(0)} + o(1)\right) \cdot (\mathrm{adj}(A(0)) + \mathrm{adj}(A)'(0) \cdot \lambda + o(\lambda))\right.$$
$$\left. + \frac{1}{\det(A)'(0)} \cdot \mathrm{adj}(A(0))\right) = \frac{1}{\lambda}\left(-\frac{\mathrm{adj}(A)'(0)}{\det(A)'(0)} \cdot \lambda + o(\lambda)\right)$$
$$= -\frac{\mathrm{adj}(A)'(0)}{\det(A)'(0)} + o(\lambda).$$

Thus, $\lim_{\lambda \to 0} \frac{1}{\lambda}(B(\lambda) - B(0))$ does exist. The same holds for higher derivations. Therefore, $g : [0,1] \to \mathrm{Sym}_2(V^*)$ is smooth and this shows the claim. $\square$

Remark 2.4.6 If we now want to transfer this to our theory, we just have to show that we can use the same arguments for every TM_p, $p \in M$, in differentiable dependence on $p \in M$. The calculations are similar to the ones above. Therefore, we define:

$$C^+ = \{s \in \Gamma(M; T^{(2,0)}M) : s_p \geq 0, \dim(\mathrm{Deg}(s_p)) = 1\}$$

and

$$TC_s^+ = \{w \in \Gamma(M; T^{(2,0)}M) : w_p(v,v) = 0, \forall v \in \mathrm{Deg}(s_p)\setminus\{0\}\}.$$

We then get the following result:

Theorem 2.4.7 *Let M be a four-dimensional connected smooth manifold. Let furthermore $g : (0,1] \to \Gamma(M; T^{(0,2)}M)$, $\lambda \mapsto g(\lambda)$ and $h : (0,1] \to \Gamma(M; T^{(2,0)}M)$, $\lambda \mapsto h(\lambda)$. The metrics $g(\lambda)$ and $h(\lambda)$ satisfy the first two conditions of the frame theory for every $\lambda \in (0,1]$. Then it is equivalent:*

(i) The metrics $g(\lambda)$ and $h(\lambda)$ have a limit for $\lambda = 0$ and these limits satisfy the first two conditions of the frame theory for $\lambda = 0$.

(ii) The limit

$$h(0) := \lim_{\lambda \to 0} h(\lambda) \in \Gamma(M; T^{(2,0)}M)$$

exists and $\mathrm{rk}(h(0)) = 3$ and $\mathrm{ind}(h(0)) = 0$. Furthermore,

$$h'(0) \notin TC_{h(0)}^+.$$

2.5 Static and spherically symmetric spacetimes

In this section we want to show that a static and spherically symmetric spacetime has a Newtonian limit. This means that we have to show that it can be extended to a family of Ehlers spacetimes and that this family has a Newtonian limit. But we first have to introduce some concepts.

Definition 2.5.1 Let M be a manifold, $g \in \Gamma(M, T^{(p,q)}M)$ a tensor field, $p, q \in \mathbb{N}_0$, and $X \in \mathfrak{X}(M)$. We then define the *Lie derivation* $\mathcal{L}$ of g with respect to X by

$$\mathcal{L} : \mathfrak{X}(M) \times \Gamma(M, T^{(p,q)}M) \to \Gamma(M, T^{(p,q)}M),$$
$$(X, g) \mapsto \mathcal{L}_X g := \left.\frac{d}{dt}\right|_{t=0} (\varphi^t)^*(g),$$

where $\varphi = (\varphi^t)$ is the flow of the vector field X.

2.5. Static and spherically symmetric spacetimes

Remark 2.5.2 Usually we will treat $(0,2)-$ tensors, so $g \in \Gamma(M, T^{(0,2)}M)$. For $p \in M$ and $\xi_1, \xi_2 \in TM_p$

$$(\mathcal{L}_X g)_p(\xi_1, \xi_2) := \left.\frac{d}{dt}\right|_{t=0} g_{\varphi^t(p)}(\varphi^t_* \xi_1, \varphi^t_* \xi_2)$$

holds. Now, if $\varphi^t : D_t \to D_{-t}$, $D_t \subseteq M$, $p \in D_t$, is a 1-parameter group of isometries, which means that $(\varphi^t)^* g = g$ holds, for all $t \in \mathbb{R}$, then we have for the corresponding vector field $X \in \mathfrak{X}(M)$, $X = \left.\frac{d}{dt}\right|_{t=0} \varphi^t$:

$$\mathcal{L}_X g = \left.\frac{d}{dt}\right|_{t=0} (\varphi^t)^* g = \left.\frac{d}{dt}\right|_{t=0} g = 0.$$

But, on the other hand:

Theorem 2.5.3 Let $X \in \mathfrak{X}(M)$ with $\mathcal{L}_X g = 0$ and let $\varphi = (\varphi^t)$ be the flow belonging to X. Then (φ^t) is a 1-parameter group of isometries.

Proof. For this proof we suppose that X is a complete vector field. If X is not complete we can use the same proof for the neighbourhood D_t of p, $t \in \mathbb{R}$, $p \in D_t$.
The following holds:

$$\frac{d}{dt}(\varphi^t)^* g = \left.\frac{d}{ds}\right|_{s=0} (\varphi^{t+s})^*(g) = \left.\frac{d}{ds}\right|_{s=0} (\varphi^s)^*((\varphi^t)^* g) = \mathcal{L}_X((\varphi^t)^* g).$$

In general, if Φ is a diffeomorphism on M, $\Phi : M \to M$, $g \in \Gamma(M, T^{(0,2)}M)$, $p \in M$ and $\xi_1, \xi_2 \in TM_p$, then we have:

$$\mathcal{L}_X(\Phi^* g)_p(\xi_1, \xi_2) = \left.\frac{d}{dt}\right|_{t=0} (\Phi^* g)_{\varphi^t(p)}(\varphi^t_* \xi_1, \varphi^t_* \xi_2) = \left.\frac{d}{dt}\right|_{t=0} g_{\Phi \circ \varphi^t(p)}(\Phi_* \circ \varphi^t_*(\xi_1), \Phi_* \circ \varphi^t_*(\xi_2))$$

and

$$(\Phi^*(\mathcal{L}_X g))_p(\xi_1, \xi_2) = (\mathcal{L}_X g)_{\Phi(p)}(\Phi_* \xi_1, \Phi_* \xi_2) = \left.\frac{d}{dt}\right|_{t=0} g_{\varphi^t(p) \circ \Phi}(\varphi^t_* \circ \Phi_*(\xi_1), \varphi^t_* \circ \Phi_*(\xi_2)).$$

Thus, if $\Phi \circ \varphi^t = \varphi^t \circ \Phi$ holds for all $t \in \mathbb{R}$, it follows that

$$\mathcal{L}_X(\Phi^* g) = \Phi^*(\mathcal{L}_X g).$$

But for all $s, t \in \mathbb{R}$

$$\varphi^t \circ \varphi^s = \varphi^{t+s} = \varphi^s \circ \varphi^t$$

holds and therefore it follows for all $t \in \mathbb{R}$:

$$\frac{d}{dt}(\varphi^t)^* g = \mathcal{L}_X((\varphi^t)^* g) = (\varphi^t)^*(\mathcal{L}_X g) = 0.$$

Thus,

$$(\varphi^t)^*(g) = (\varphi^0)^* g = g$$

2.5. Static and spherically symmetric spacetimes

for all $t \in \mathbb{R}$, which shows the claim. □

Definition 2.5.4 A vector field $X \in \mathfrak{X}(M)$ with $\mathcal{L}_X g = 0$ is called *Killing vector field*.

Definition 2.5.5 Let (M, g) be a spacetime. A timelike future-oriented unit vector field U is called an *observer*.

Definition 2.5.6 We call an observer U *irrotational* if for all vector fields X, Y which are orthogonal to U
$$\langle \nabla_Y U, X \rangle - \langle \nabla_X U, Y \rangle = 0$$
holds.

Theorem 2.5.7 *An observer U is irrotational if and only if there is a hypersurface $N \subset M$ which is orthogonal to U in every $p \in M$, that means that there is a three-dimensional submanifold $N \subset M$ with $p \in N$ and $TN_q \perp U_q$ for $q \in N$.*

Proof. The proof follows from the theorem of Frobenius (see (1.4.3)), according to which there is such an orthogonal hypersurface N in every $p \in U$ if and only if the condition
$$[X, Y] \perp U, \quad \forall X, Y \in \mathfrak{X}(M) \text{ with } X \perp U, Y \perp U$$
holds. But this is equivalent to the condition of irrotationality, since
$$\langle \nabla_Y U, X \rangle - \langle \nabla_X U, Y \rangle = \nabla_Y \langle U, X \rangle - \langle U, \nabla_Y X \rangle - \nabla_X \langle U, Y \rangle + \langle U, \nabla_X Y \rangle$$
$$= 0 + \langle U, \nabla_X Y - \nabla_Y X \rangle = \langle U, [X, Y] \rangle.$$

□

Definition 2.5.8 1. A spacetime (M, g) is called *stationary* if there is a global timelike Killing vector field V on M.

2. If, furthermore, the corresponding observer $U = \frac{V}{\|V\|}$ is irrotational, the spacetime is called *static*.

Theorem 2.5.9 *Let the spacetime (M, g) be static and $p \in M$. Then there is a chart $\varphi = (t, x^1, x^2, x^3) : U \to I \times \mathbb{B}^3$, $p \in U$, and there are maps $f : \mathbb{B}^3 \to (0, \infty)$ and $h : \mathbb{B}^3 \to Pos_3(\mathbb{R})$ so that for the representation of g with respect to φ*
$$ds^2 = -f^2 dt^2 + d\sigma^2$$
holds, where $h = d\sigma^2$ is a Riemannian metric on $\mathbb{B}^3$.

Proof. Let $p \in M$, $N \subseteq M$ be the integral submanifold with $p \in N$, X the timelike Killing vector field and (ψ^t) the (local) flow of isometries belonging to X. We then consider $\Phi : \mathbb{R} \times N \to M$,
$$\Phi(t, q) := \psi^t(q),$$

2.5. Static and spherically symmetric spacetimes

(if (ψ^t) is not global, choose $\Phi : I \times V \to U$ with $I \subseteq \mathbb{R}$ an open interval and $0 \in I$, as well as $V \subseteq N$ an open neighbourhood of p and $U \subseteq M$ an open neighbourhood of p). Then:

$$D\Phi_{(0,p)}\left(\frac{\partial}{\partial t}\right) = X_p, \quad D\Phi|_{TN_p} = \mathrm{id}_{TN_p}, \quad TN_p = X_p^\perp,$$

thus, $D\Phi_{(0,p)}$ is an isomorphism and $\Phi|_{I \times V} : I \times V \to U$ a diffeomorphism for $I \subseteq \mathbb{R}$ an open interval and $0 \in I$, as well as $V \subseteq N$ an open neighbourhood of p and $U \subseteq M$ an open neighbourhood of p. After a possible diminution of V and by using the chart of N in a neighbourhood of p we can suppose that $V = \mathbb{B}^3$ without restriction.
Φ then maps the coordinate planes $\{t = t_0\} \subseteq I \times \mathbb{B}^3$ to the leaves $\psi^t(N) \cap U$. We now define $\varphi := \Phi^{-1}$. Then for the Lorenzian metric g with respect to φ

$$g_{00}(t,x) = g(\partial_t, \partial_t) = g(X,X) =: -f^2(t,x)$$

holds, since $\partial_t = X$, if ∂_t denotes the coordinate vector field on U. But as X is a Killing vector field its length stays constant along the integral curves due to the fact that

$$X(\psi^t(q)) = D\psi^t_q(X_q).$$

Therefore, f does not depend on t. Furthermore, we get

$$g_{0i}(t,x) = g\left(\partial_t, \frac{\partial}{\partial x^i}\right) = g\left(X, \frac{\partial}{\partial x^i}\right) = 0,$$

for $i = 1, 2, 3$, since X_q is always orthogonal to N_q.
Finally, ψ^t maps N to $\psi^t(N)$ so that the map is an isometry. Therefore,

$$g_{ij}(t,x) = g\left(\frac{\partial}{\partial x^i}, \frac{\partial}{\partial x^j}\right) = g\left(\frac{\partial}{\partial x^i}, \frac{\partial}{\partial x^j}\right)_{\psi^t(\Phi(0,x))} = g\left(\frac{\partial}{\partial x^i}, \frac{\partial}{\partial x^j}\right)_{\Phi(0,x)} = g_{ij}(0,x)$$

does not depend on t, for $1 \leq i, j \leq 3$. If we now define $h : \mathbb{B}^3 \to \mathrm{Pos}(\mathbb{R})$,

$$h_{ij}(x) := g_{ij}(0,x),$$

we finally get that

$$ds^2 = -f^2 dt^2 + d\sigma^2$$

holds with $h =: d\sigma^2$ and $g =: ds^2$. □

Remark 2.5.10 Theorem (2.5.9) only provides a local product structure, of course. But in the following we want to have a global structure so that

$$M \cong I \times M^3.$$

But this is clearly not the case for every static spacetime. We therefore have to require that

$$\Phi : I \times M^3 \to M, \; \Phi(t,p) = \psi^t(p),$$

is a diffeomorphism, where (ψ^t) is the flow belonging to the Killing vector field X. Then M has the desired form. If we talk about a static spacetime in the following, we always mean that this addition requirement is satisfied and thus the manifold is of the desired form.

Definition 2.5.11 A spacetime (M,g) is called *spherically symmetric*, if $SO(3)$ operates as a group of isometries so that the orbits are diffeomorphic to $\mathbb{S}^2$.

Remark 2.5.12 Sometimes, a spacetime (M,g) is called spherically symmetric, if $SO(3)$ operates as a group of isometries so that the orbits are two-dimensional spacelike surfaces. If we then consider the operation

$$\tau : SO(3) \times M \to M,$$

with $\dim M = 3$, we get for the isotropy group H:

$$B \cong SO(3)/H,$$

where B denotes the orbit, thus $\dim(B) = 2$. Furthermore, $\dim(SO(3)) = 3$, therefore it follows that $\dim(H) = 1$. But we know that the only one-dimensional subgroups of $SO(3)$ are conjugates of $SO(2)$ and $O(2)$. The first case provides

$$SO(3)/SO(2) \cong \mathbb{S}^2,$$

which also holds according to our definition. In the other case we get

$$SO(3)/O(2) \cong \mathbb{P}^2,$$

which is not the result we want to have. But if you do not want to use the stronger definition of spherically symmetric spacetimes you can also require the orbits to be simply connected or the manifold to be oriented. This also excludes the second case and the orbits are $\mathbb{S}^2$ indeed.

Theorem 2.5.13 *Let (M,g) be a spacetime which is static and spherically symmetric. Furthermore, let M be simply connected. Then,*

$$M \cong I \times J \times \mathbb{S}^2$$

as manifolds, where $I \subseteq \mathbb{R}$ and $J \subseteq \mathbb{R}$ are open intervals.

Proof. As mentioned in (2.5.10), we suppose that $M = I \times M^3$ due to the fact that M is static. We now define

$$N := M^3/SO(3).$$

As $SO(3)$ operates as a group of isometries so that all the orbits are diffeomorphic to $\mathbb{S}^2$, all the isometry groups H_p are conjugates. We can therefore use the slice theorem (see [3], chap. 2, §5) which guarantees that N is a smooth manifold. Furthermore, it provides that M is an $\mathbb{S}^2$–bundle. Therefore, $M^3 \to M^3/SO(3) = N$ is a fibration with fibre $\mathbb{S}^2$. As $N = M^3/SO(3)$, we know that $\dim N = 1$ and N is either an open interval or $N = \mathbb{S}^1$. But we required that M is simply connected, and thus, M^3 is also simply connected. Therefore, the long homotopy sequence

$$\begin{array}{cccccc}
\pi_1(M^3) & \to & \pi_1(N) & \to & \pi_0(\mathbb{S}^2) & \to & \pi_0(M^3) \\
= (1) & & & & = (0) & & = (0)
\end{array}$$

for $\mathbb{S}^2 \to M^3 \to N$ is exact. Thus, $\pi_1(N) = (1)$, which means that N is simply connected and therefore, $N \cong J$, where $J \subseteq \mathbb{R}$ is an open interval.

Thus, M is a trivial bundle since J is simply connected, which means that

$$M \cong I \times J \times \mathbb{S}^2.$$

□

Now we know the structure of the manifold and we also know that

$$ds^2 = -f^2 dt^2 + d\sigma^2$$

holds. Now we deal with the metric $h := d\sigma^2$. We have:

Lemma 2.5.14 *Let g be a metric on $\mathbb{S}^2$ which is invariable under $SO(3)$-transformations (this means that for all $\Phi \in SO(3)$, $\Phi : \mathbb{S}^2 \to \mathbb{S}^2$, $\Phi^* g = g$ holds). Then there is a constant $c > 0$ so that $g = c g_{sph}$, where g_{sph} is the standard metric on $\mathbb{S}^2$.*

Proof. Let N be the north pole of $\mathbb{S}^2$, $T\mathbb{S}_N^2$ the tangent space in the north pole and g_N the metric in the north pole. Since $SO(3)$ operates as a group of isometries on $\mathbb{S}^2$, the metric also has to be invariable under rotations around the x^3-axis. The only metrics that satisfy this condition are of the form

$$g_N = c \langle \_, \_ \rangle_{\text{sph}, N},$$

here $T\mathbb{S}_N^2$ is identified with $\mathbb{R}^2$. Now this holds since the symmetric bilinear form $B : \mathbb{R}^n \times \mathbb{R}^n \to \mathbb{R}$ is defined by its corresponding quadratic form $q_B : \mathbb{R}^n \to \mathbb{R}$, $q_B(v) := B(v,v)$. But, on the other hand, this quadratic form is fixed by its values on $\mathbb{S}^{n-1} \subseteq \mathbb{R}^n$, since $q_B(\lambda v) = \lambda^2 q_B(v)$. If now B is invariable under $SO(n)$, $q_B\big|_{\mathbb{S}^{n-1}}$ is also invariable and therefore constant, as $SO(n)$ operates transitively on $\mathbb{S}^{n-1}$. Thus, $q_B\big|_{\mathbb{S}^{n-1}} = c$. But this means that $B = c \langle \_, \_ \rangle_{\text{std}}$.

However, c could vary from one point to another. But, since $SO(3)$ operates as a group of isometries on $\mathbb{S}^2$ and furthermore transitively, it follows from $\Phi^* g_{sph} = g_{sph}$ and $\Phi^* g = g$, $\forall \Phi \in SO(3)$ that $\Phi^* c = c$ and therefore for every $p \in \mathbb{S}^2$:

$$g_p = c \langle \_, \_ \rangle_{\text{sph}, p}.$$

Thus, all in all,
$$g = c\langle \cdot, \cdot \rangle_{\mathrm{sph}} = c g_{\mathrm{sph}}$$
holds. □

Lemma 2.5.15 *Let g be a Riemannian metric on $M = \mathbb{R}_+ \times \mathbb{S}^2$ and let $SO(3)$ operate as a group of isometries on $\mathbb{S}^2$. Then:*
$$g_{(\rho,\xi)}(\partial_\rho, Y) = 0, \quad \forall Y \in \mathfrak{X}(\mathbb{S}^2),$$
where ρ is a "proper coordinate" on $\mathbb{R}_+$ and ξ a "generalized coordinate" on $\mathbb{S}^2$.

Proof. Let $\xi \in \mathbb{S}^2$ be arbitrary and $\eta \in T\mathbb{S}^2_\xi \subseteq \mathbb{R}^3$, thus $\xi \perp \eta$. Then there is an $A \in SO(3)$ with $A(\xi) = \xi$ and $A(\eta) = -\eta$ (for example, reflection at the plane $\mathrm{span}(\xi, \zeta)$, if $\zeta \perp \xi$ and $\zeta \perp \eta$). Furthermore, let
$$\Phi : M \to M, \quad (\rho, \xi) \mapsto (\rho, A\xi).$$
Then $\Phi^* g = g$ as well as $\Phi(\rho, \xi) = (\rho, \xi)$ hold and it follows:
$$g_{(\rho,\xi)}(\partial_\rho, \eta) = (\Phi^* g)_{(\rho,\xi)}(\partial_\rho, \eta) = g_{(\rho,\xi)}(\Phi_* \partial_\rho, \Phi_* \eta) = g_{(\rho,\xi)}(\partial_\rho, -\eta) = -g_{(\rho,\xi)}(\partial_\rho, \eta)$$
$$\Rightarrow g_{(\rho,\xi)}(\partial_\rho, \eta) = 0,$$
$\forall \eta \in T\mathbb{S}^2_\xi$. Since $\xi \in \mathbb{S}^2$ is arbitrary, we have
$$g_{(\rho,\xi)}(\partial_\rho, Y) = 0, \quad \forall Y \in \mathfrak{X}(\mathbb{S}^2).$$
□

We now want to show that a static and spherically symmetric spacetime has a genuine Newtonian limit. We therefore have to consider the following:

Theorem 2.5.16 *Let (M, g) be a static and spherically symmetric spacetime. Then*
$$M \cong I \times J \times \mathbb{S}^2$$
as manifold and the diffeomorphism $\Phi : M \to I \times J \times \mathbb{S}^2$ provides coordinates with respect to which
$$ds^2 = A(\rho) dt^2 + B(\rho) d\rho^2 + C(\rho) d\sigma^2$$
holds, with $g = ds^2$ and $d\sigma^2$ the standard metric on $\mathbb{S}^2$.

Proof. First we have to consider that we already know from (2.5.13) and (2.5.9) that the manifold is of the form $M = I \times J \times \mathbb{S}^2$ and that there are functions $f : J \times \mathbb{S}^2 \to (0, \infty)$ and $h : J \times \mathbb{S}^2 \to \mathrm{Pos}_3(\mathbb{R})$ so that for g
$$ds^2 = -f^2 dt^2 + d\tau^2$$

holds, where $h = d\tau^2$ is a Riemannian metric on $J \times \mathbb{S}^2$.

Thus we now have to show that the Riemannian metric h has diagonal form and that all components of g only depend on one coordinate. Therefore, let (t, ρ, ξ) be the coordinates for M. Here (t, ρ) are "regular coordinates", $\xi \in \mathbb{S}^2$ is a "generalized coordinate" on $\mathbb{S}^2$.

First we consider the Riemannian metric on $J \times \mathbb{S}^2$. According to lemma (2.5.14) and lemma (2.5.15) we already know that the metric is diagonal and has the form

$$g(\rho, \xi) = B(\rho, \xi) d\rho^2 + C(\rho, \xi) d\sigma^2,$$

where $d\sigma^2$ is the standard metric on $\mathbb{S}^2$.

Therefore, the metric g on M has the form

$$g(\rho, \xi) = A(\rho, \xi) dt^2 + B(\rho, \xi) d\rho^2 + C(\rho, \xi) d\sigma^2.$$

But since $SO(3)$ operates as a group of isometries,

$$g(\rho, \Phi\xi) = g(\rho, \xi)$$

holds $\forall \Phi \in SO(3)$. Thus it already follows that A, B and C do not depend on ξ. But this means the g is of the form

$$ds^2 = A(\rho) dt^2 + B(\rho) d\rho^2 + C(\rho) d\sigma^2.$$

This shows the claim. □

Theorem 2.5.17 *Let (M, g) be a static and spherically symmetric spacetime. Then we can find an extension of this spacetime so that the resulting bilinear forms $g(\lambda)$ and $h(\lambda)$ have limits for $\lambda = 0$ which satisfy the axioms of the frame theory for $\lambda = 0$.*

Proof. We already know that for the static and spherically symmetric manifold M and its metric g

$$M = I \times J \times \mathbb{S}^2$$

and

$$ds^2 = A(\rho) dt^2 + B(\rho) d\rho^2 + C(\rho) d\sigma^2$$

hold. Of course, the metric is a Lorentzian metric, therefore it has index 1 and is of full rank. Since t is the time coordinate, it follows that $A(\rho)$ is negative, while the other functions are positive. We now choose the following diffeomorphism for $\lambda > 0$:

$$\Phi : I \to \tilde{I} := \Phi(I), \quad t \mapsto \frac{1}{\sqrt{\lambda}} t.$$

We therefore now get a metric depending on λ which is of the form

$$\widetilde{ds}^2 = \frac{1}{\lambda} A(\rho) dt^2 + B(\rho) d\rho^2 + C(\rho) d\sigma^2.$$

In order to get the metric $g(\lambda)$, we now have to multiply $d\tilde{s}^2$ with $-\lambda$, which is not zero for $\lambda > 0$, and we get:
$$ds^2 = -A(\rho)dt^2 - \lambda B(\rho)d\rho^2 - \lambda C(\rho)d\sigma^2.$$

Therefore, $g(\lambda)$ has index 3 (as we multiply a Lorentzian metric with a negative number) and is still of full rank.

Now we want to find the metric $h(\lambda)$. We again start with
$$d\tilde{s}^2 = \frac{1}{\lambda}A(\rho)dt^2 + B(\rho)d\rho^2 + C(\rho)d\sigma^2.$$

If we now invert this metric we get for $h(\lambda)$:
$$\left(\frac{\partial}{\partial s}\right)^2 = \frac{\lambda}{A(\rho)}\left(\frac{\partial}{\partial t}\right)^2 + \frac{1}{B(\rho)}\left(\frac{\partial}{\partial \rho}\right)^2 + \frac{1}{C(\rho)}\left(\frac{\partial}{\partial \sigma}\right)^2.$$

Thus, $h(\lambda)$ is of full rank and has index 1 due to the fact that $\lambda > 0$. For the corresponding transitions from tangent space to the cotangent space and vice versa, $\varphi_p(\lambda) \circ \psi_p(\lambda) = -\lambda \operatorname{id}_{TM_p^*}$ holds, of course. Thus, the extended metrics satisfy the first two axioms of the frame theory for $\lambda > 0$.

Now we have to consider the limits. For $\lambda \to 0$ we then have for $g(0)$:
$$ds^2 = -A(\rho)dt^2.$$

Therefore, $g(0)$ has rank 1 and index 0, since $-A(\rho)$ is positive. Moreover, it is clear that the limit exists since λ only appears as a factor in the last two components of the metric.

Now we talk about the limit $h(0)$. It has the form
$$\left(\frac{\partial}{\partial s}\right)^2 = \frac{1}{B(\rho)}\left(\frac{\partial}{\partial \rho}\right)^2 + \frac{1}{C(\rho)}\left(\frac{\partial}{\partial \sigma}\right)^2.$$

Thus, $h(0)$ has rank 3 as well as index 0, since $B(\rho)$ and $C(\rho)$ are positive.

In this case $\varphi_p(0) \circ \psi_p(0) = 0$ holds for the corresponding maps as well. Therefore, the limits also satisfy the first two axioms of the frame theory and we have found a possible extension of the metric. □

Remark 2.5.18 a) First, we have to comment on the extension that we have chosen in this theorem. We chose the simplest possibility to extend the metrics. In this extension the functions $A(\rho)$, $B(\rho)$ and $C(\rho)$ do not depend on λ, so the extension here is just $A_\lambda(\rho) := A(\rho)$, $B_\lambda(\rho) := B(\rho)$ and $C_\lambda(\rho) := C(\rho)$. If you consider the extensions of the Minkowski, Schwarzschild or Kerr spacetimes (see section 2.3), you realize that these extensions are not so simple, as the functions there also depend on λ. As we will soon see, we have to use another extension in order to make sure that not only the metrics but also the connections have a limit for $\lambda = 0$.

b) According to section (2.4) the metrics $g(\lambda)$ have a limit if the metrics $h(\lambda)$ have one and

$h'(0) \notin TC^+_{h(0)}$. But this is the case for our extension due to the fact that

$$\frac{d}{d\lambda}\bigg|_{\lambda=0} \frac{\lambda}{A(\rho)} = \frac{A(\rho) - \lambda \cdot 0}{(A(\rho))^2}\bigg|_{\lambda=0} = \frac{1}{A(\rho)} \neq 0.$$

Theorem 2.5.19 *Let (M,g) be a static and spherically symmetric spacetime. Then there is an extension of (M,g) so that the connections $\nabla(\lambda)$ belonging to $g(\lambda)$ have a limit for $\lambda = 0$.*

Proof. We already know that there is an extension of a static and spherically symmetric spacetime so that the metrics have limits which satisfy the conditions of the frame theory. In order to guarantee that the connection also has a limit we have to use an extension of the form

$$ds^2 = -A_\lambda(\rho)dt^2 - \lambda B_\lambda(\rho)d\rho^2 - \lambda C_\lambda(\rho)d\sigma^2$$

for g and

$$\left(\frac{\partial}{\partial s}\right)^2 = \frac{\lambda}{A_\lambda(\rho)}\left(\frac{\partial}{\partial t}\right)^2 + \frac{1}{B_\lambda(\rho)}\left(\frac{\partial}{\partial \rho}\right)^2 + \frac{1}{C_\lambda(\rho)}\left(\frac{\partial}{\partial \sigma}\right)^2$$

for h. In the following we will discuss the functions $A_\lambda(\rho)$, $B_\lambda(\rho)$ and $C_\lambda(\rho)$ more closely. We now first consider the components of the extended connection. As for $\lambda > 0$ the connection $\nabla(\lambda)$ is the Levi-Civita connection of $g(\lambda)$ we can get the components of the connection from the components of the extended metric. We therefore consider the metric $g(\lambda)$ with respect to the coordinates $(t, \rho, \vartheta, \varphi)$. We then get (remember that the standard metric $d\sigma^2$ on $\mathbb{S}^2$ in spherical coordinates $(\vartheta, \varphi) \in (0, \pi) \times (0, 2\pi)$ is of the form $d\sigma^2 = d\vartheta^2 + \sin^2\vartheta d\varphi^2$):

$$(g_{ij}(\lambda)) = \mathrm{diag}(-A_\lambda(\rho), -\lambda B_\lambda(\rho), -\lambda C_\lambda(\rho), -\lambda \cdot \sin^2\vartheta C_\lambda(\rho)),$$

as well as

$$(g^{ij}(\lambda)) = \mathrm{diag}\left(\frac{1}{-A_\lambda(\rho)}, \frac{1}{-\lambda B_\lambda(\rho)}, \frac{1}{-\lambda C_\lambda(\rho)}, \frac{1}{-\lambda \cdot \sin^2\vartheta C_\lambda(\rho)}\right).$$

Due to the diagonal form of the metric the following holds for the components of the connection:

$$\Gamma^i_{ii} = \frac{1}{2}g^{ii}\partial_i g_{ii}, \quad i = 0, ..., 3;$$

$$\Gamma^k_{ii} = -\frac{1}{2}g^{kk}\partial_k g_{ii}, \quad i, k = 0, ..., 3; \quad i \neq k;$$

$$\Gamma^j_{ij} = \frac{1}{2}g^{jj}\partial_i g_{jj}, \quad i, j = 0, ..., 3; \quad i \neq j;$$

$$\Gamma^k_{ij} = 0 \text{ otherwise.}$$

Furthermore, we have to bear in mind that the components of the metric only depend on ρ and ϑ. Therefore, the derivations with respect to t and φ vanish immediately. The components of

2.5. Static and spherically symmetric spacetimes

the connection that are not zero are:

$$\Gamma^1_{00} = -\frac{1}{2}\frac{1}{-\lambda B_\lambda(\rho)} \cdot \partial_1(-A_\lambda(\rho)) = -\frac{1}{2}\frac{\partial_1(A_\lambda(\rho))}{\lambda B_\lambda(\rho)}$$

$$\Gamma^1_{11} = \frac{1}{2}\frac{1}{-\lambda B_\lambda(\rho)} \cdot \partial_1(-\lambda B_\lambda(\rho)) = \frac{1}{2}\frac{\partial_1(B_\lambda(\rho))}{B_\lambda(\rho)}$$

$$\Gamma^1_{22} = -\frac{1}{2}\frac{1}{-\lambda B_\lambda(\rho)} \cdot \partial_1(-\lambda C_\lambda(\rho)) = -\frac{1}{2}\frac{\partial_1(C_\lambda(\rho))}{B_\lambda(\rho)}$$

$$\Gamma^1_{33} = -\frac{1}{2}\frac{1}{-\lambda B_\lambda(\rho)} \cdot \partial_1(-\lambda \sin^2\vartheta \cdot C_\lambda(\rho)) = -\frac{1}{2}\frac{\partial_1(\sin^2\vartheta \cdot C_\lambda(\rho))}{B_\lambda(\rho)}$$

$$\Gamma^2_{33} = -\frac{1}{2}\frac{1}{-\lambda C_\lambda(\rho)} \cdot \partial_2(-\lambda \sin^2\vartheta \cdot C_\lambda(\rho)) = -\frac{1}{2} \cdot 2\sin\vartheta \cdot \cos\vartheta = -\sin\vartheta \cdot \cos\vartheta$$

$$\Gamma^0_{10} = \frac{1}{2}\frac{1}{-A_\lambda(\rho)} \cdot \partial_1(-A_\lambda(\rho)) = \frac{1}{2}\frac{\partial_1(A_\lambda(\rho))}{A_\lambda(\rho)}$$

$$\Gamma^2_{12} = \frac{1}{2}\frac{1}{-\lambda C_\lambda(\rho)} \cdot \partial_1(-\lambda C_\lambda(\rho)) = \frac{1}{2}\frac{\partial_1(C_\lambda(\rho))}{C_\lambda(\rho)}$$

$$\Gamma^3_{13} = \frac{1}{2}\frac{1}{-\lambda \sin^2\vartheta \cdot C_\lambda(\rho)} \cdot \partial_1(-\lambda \sin^2\vartheta \cdot C_\lambda(\rho)) = \frac{1}{2}\frac{\partial_1(C_\lambda(\rho))}{C_\lambda(\rho)}$$

$$\Gamma^3_{23} = \frac{1}{2}\frac{1}{-\lambda \sin^2\vartheta \cdot C_\lambda(\rho)} \cdot \partial_2(-\lambda \sin^2\vartheta \cdot C_\lambda(\rho)) = \frac{1}{2}\frac{2\sin\vartheta \cdot \cos\vartheta}{\sin^2\vartheta} = \frac{\cos\vartheta}{\sin\vartheta}.$$

If we want the limit to exist, we have to require for the extension:

$$\lim_{\lambda \to 0} \frac{\partial_1(A_\lambda(\rho))}{\lambda B_\lambda(\rho)} \text{ does exist;}$$

$$\lim_{\lambda \to 0}(A_\lambda(\rho)) \neq 0;$$

$$\lim_{\lambda \to 0}(B_\lambda(\rho)) \neq 0;$$

$$\lim_{\lambda \to 0}(C_\lambda(\rho)) \neq 0.$$

If we therefore require that

$$A_\lambda(\rho) := 1 + (A(\rho) - 1)\lambda$$
$$B_\lambda(\rho) := B(\rho)$$
$$C_\lambda(\rho) := C(\rho)$$

for $\lambda \in [0,1]$ we have found an extension which ensures the existence of the limit. First we have to make sure that this is an extension of the metric g, but this is clearly the case since

$$A_1(\rho) = 1 + A(\rho) - 1 = A(\rho).$$

Furthermore, the conditions from above are satisfied since

$$\lim_{\lambda \to 0} \frac{\partial_1(A_\lambda(\rho))}{\lambda B_\lambda(\rho)} = \lim_{\lambda \to 0} \frac{\partial_1(1+(A(\rho)-1)\lambda)}{\lambda B(\rho)} = \lim_{\lambda \to 0} \frac{\lambda \partial_1(A(\rho))}{\lambda B(\rho)} = \lim_{\lambda \to 0} \frac{\partial_1(A(\rho))}{B(\rho)}$$
$$= \frac{\partial_1(A(\rho))}{B(\rho)} \text{ exists, since } B(\rho) \neq 0;$$
$$\lim_{\lambda \to 0}(A_\lambda(\rho)) = \lim_{\lambda \to 0}(1+(A(\rho)-1)\lambda) = 1;$$
$$\lim_{\lambda \to 0}(B_\lambda(\rho)) = \lim_{\lambda \to 0}(B(\rho)) = B(\rho) \neq 0;$$
$$\lim_{\lambda \to 0}(C_\lambda(\rho)) = \lim_{\lambda \to 0}(C(\rho)) = C(\rho) \neq 0.$$

Therefore, the connections $\nabla(\lambda)$ have a limit for $\lambda = 0$. $\square$

Remark 2.5.20 Of course, we do not only want to have a limit connection but we also want it to satisfy the corresponding axioms of the frame theory for $\lambda = 0$. We therefore have to guarantee that the connections converge in a $\mathcal{C}^2$-sense at least. Then the limits of the first and second derivations also exist and the axioms five and six follow immediately. Moreover, $\nabla g = 0$ as well as $\nabla h = 0$ also follow since, in this case, forming of limits and derivations can be exchanged. But this means that you just have to derivate the Christoffel symbols from above and then adapt the functions $A_\lambda(\rho)$, $B_\lambda(\rho)$ and $C_\lambda(\rho)$ as we did in the proof of theorem (2.5.19).

Using the same approach we can also make sure that the matter tensor $T(\lambda)$ (remember that $T(\lambda)$ is fixed by $g(\lambda)$ for $\lambda \in (0,1]$) has a limit for $\lambda = 0$ and that this limit satisfies the corresponding axioms of the frame theory for $\lambda = 0$. But this now shows that we get the following theorem:

Theorem 2.5.21 *Let (M, g) be a static and spherically symmetric spacetime. Then there is an extension $g(\lambda)$ of the metric g so that $g(\lambda)$ and the corresponding objects $h(\lambda)$, $\nabla(\lambda)$ and $T(\lambda)$ for $\lambda \in (0, 1]$ have a limit for $\lambda = 0$. Furthermore, these limits satisfy the axioms of the frame theory.*
Thus, the extended family of Ehlers spacetimes $(g(\lambda), h(\lambda), \nabla(\lambda), T(\lambda), \lambda)$ has a quasi-Newtonian limit.

We even get the following result:

Theorem 2.5.22 *Let (M, g) be a static and spherically symmetric spacetime. Then there is an extension $g(\lambda)$ of the metric g so that $g(\lambda)$ and the corresponding objects $h(\lambda)$, $\nabla(\lambda)$ and $T(\lambda)$ for $\lambda \in (0, 1]$ have a limit for $\lambda = 0$. These limits satisfy the axioms of the frame theory. Furthermore, $v = 0$ in the corresponding coordinates.*
Thus, the extended family of Ehlers spacetimes $(g(\lambda), h(\lambda), \nabla(\lambda), T(\lambda), \lambda)$ has a genuine Newtonian limit.

Proof. As we already know that the family of Ehlers spacetimes has a quasi-Newtonian limit we just have to discuss the condition $v = 0$. We already know that the metrics $g(\lambda)$ have diagonal form for $\lambda > 0$. Then we have for the Christoffel symbols:

$$\Gamma^i_{ii} = \frac{1}{2} g^{ii} \partial_i g_{ii}, \quad i = 0, ..., 3;$$

$$\Gamma^k_{ii} = -\frac{1}{2} g^{kk} \partial_k g_{ii}, \quad i, k = 0, ..., 3; \quad i \neq k;$$

$$\Gamma^j_{ij} = \frac{1}{2} g^{jj} \partial_i g_{jj}, \quad i, j = 0, ..., 3; \quad i \neq j;$$

$$\Gamma^k_{ij} = 0 \text{ otherwise.}$$

If we now define $v(\lambda)$ by $v(\lambda) = (\Gamma^3_{02}, \Gamma^1_{03}, \Gamma^2_{01})$, we already have $v(\lambda) = 0$ for $\lambda > 0$. If we consider the limit for $\lambda \to 0$ we still have $v = 0$ in case of $\lambda = 0$. But this exactly means that the limit is a genuine Newtonian one in the coordinates given above. $\square$

Chapter 3

Existence of genuine Newtonian limits

In this chapter we will discuss the question when a quasi-Newtonian limit is a genuine Newtonian one. Let therefore $M_\lambda = (M, g(\lambda), h(\lambda), \nabla(\lambda), T(\lambda), \lambda)$, $\lambda \in (0,1]$, be a family of Ehlers spacetimes and $M_0 = (M, g(0), h(0), \nabla(0), T(0))$ its quasi-Newtonian limit. According to the first chapter we know that in case of $\lambda = 0$ we can find an *adapted coordinate system* $(t,x) \in I \times U$, $I \subseteq \mathbb{R}$, $U \subseteq \mathbb{R}^3$ in a neighbourhood of every $p \in M$. In the following we suppose that the topological conditions (simply connected and geodesically complete) are satisfied. Thus the manifold is of the form

$$M \cong \mathbb{R} \times \mathbb{R}^3$$

and we get a global, adapted coordinate system $(t,x) \in \mathbb{R} \times \mathbb{R}^3$. Furthermore, we know that the admitted coordinate transformations are of the form

$$\tau : \mathbb{R} \times \mathbb{R}^3 \to \mathbb{R} \times \mathbb{R}^3,$$
$$(t,x) \mapsto (\pm t + c, S(t)x + b(t))$$

with $c \in \mathbb{R}$, $b : \mathbb{R} \to \mathbb{R}^3$ and $S : \mathbb{R} \to O(3)$.

If we now want to get a genuine Newtonian limit, we know from its definition that we have to show that there is an adapted coordinate system with respect to which

$$v = 0$$

holds, where $v = (v_1, v_2, v_3)$ and $v_i := \epsilon_{ij}^k v_k^j$, $1 \leq i,j,k \leq 3$.

We therefore first have a look at the behaviour of v in case of transformation of coordinates. This then gives us a first condition which ensures the existence of such an adapted coordinate system and therefore of a genuine Newtonian limit. Then we will discuss some conditions for the curvature tensor which guarantee that there is a genuine Newtonian limit. Eventually we will examine the asymptotically flat case which leads to the existence of a genuine Newtonian limit under certain circumstances.

3.1 Transformation of coordinates

We first have a look at the definition of v of chapter one. Let (t, x^1, x^2, x^3) be an adapted coordinate system for M. Then we have

$$\nabla_{\partial_i} B = \Gamma^j_{i0} \partial_j =: v^j_i \partial_j$$

for $1 \leq i, j \leq 3$ and $B := \frac{\partial}{\partial t}$. Therefore, we now study the behaviour of the Christoffel symbols in case of transformation of coordinates.

Let $\partial_i := \frac{\partial}{\partial x^i}$ as well as $\tilde{\partial}_i := \frac{\partial}{\partial y^i}$, $i = 0, ..., 3$, be the coordinate vector fields with regard to maps x and y respectively. With regard to these coordinate vector fields we then have:

$$\nabla_{\partial_i} \partial_j = \Gamma^k_{ij}(x) \partial_k \text{ and } \nabla_{\tilde{\partial}_i} \tilde{\partial}_j = \tilde{\Gamma}^k_{ij}(y) \tilde{\partial}_k.$$

If we now consider $y = y(x)$ as a map of x, the coordinate vector fields and 1-forms transform to

$$\tilde{\partial}_i = \frac{\partial x^i}{\partial y^j} \partial_i \text{ and } dy^k = \frac{\partial y^k}{\partial x^i} dx^i.$$

The transformed Christoffel symbols $\tilde{\Gamma}^k_{ij}$ then have the form (we omit $\circ y(x)$ in the following calculation):

$$\tilde{\Gamma}^k_{ij} = dy^k \left(\nabla_{\tilde{\partial}_i} \tilde{\partial}_j \right) = \frac{\partial y^k}{\partial x^p} dx^p \left(\nabla_{\frac{\partial x^m}{\partial y^i} \partial_m} \left(\frac{\partial x^n}{\partial y^j} \partial_n \right) \right)$$

$$= \frac{\partial y^k}{\partial x^p} \frac{\partial x^m}{\partial y^i} dx^p \left(\frac{\partial^2 x^n}{\partial y^j \partial y^l} \cdot \frac{\partial y^l}{\partial x^m} \partial_n + \frac{\partial x^n}{\partial y^j} \nabla_{\partial_m} \partial_n \right)$$

$$= \frac{\partial y^k}{\partial x^p} \frac{\partial x^m}{\partial y^i} \left(\frac{\partial^2 x^p}{\partial y^j \partial y^l} \cdot \frac{\partial y^l}{\partial x^m} + \frac{\partial x^n}{\partial y^j} dx^p \left(\Gamma^q_{mn} \partial_q \right) \right)$$

$$= \frac{\partial y^k}{\partial x^p} \frac{\partial x^m}{\partial y^i} \left(\frac{\partial^2 x^p}{\partial y^j \partial y^l} \cdot \frac{\partial y^l}{\partial x^m} + \frac{\partial x^n}{\partial y^j} \Gamma^p_{mn} \right).$$

Altogether we then have:

$$\tilde{\Gamma}^k_{ij}(y(x)) = \left(\frac{\partial y^k}{\partial x^p} \cdot \frac{\partial x^m}{\partial y^i} \circ y \left(\frac{\partial^2 x^p}{\partial y^j \partial y^l} \circ y \cdot \frac{\partial y^l}{\partial x^m} + \frac{\partial x^n}{\partial y^j} \circ y \cdot \Gamma^p_{mn} \right) \right)(x).$$

Now we know how the Christoffel symbols behave under coordinate transformation. Furthermore, the admitted coordinate transformations are given by

$$\tau : \mathbb{R} \times \mathbb{R}^3 \to \mathbb{R} \times \mathbb{R}^3, \quad (t, x) \mapsto (\pm t + c, S(t) x + b(t)),$$

where $c \in \mathbb{R}$, $S : \mathbb{R} \to O(3)$ and $b : \mathbb{R} \to \mathbb{R}^3$.

Now let $x' = (t, x^1, x^2, x^3)$ and $y' = (s, y^1, y^2, y^3)$ be adapted coordinate systems for M. Given

3.1. Transformation of coordinates

a coordinate transformation from x' to y', which means that

$$y' = \begin{pmatrix} s \\ y \end{pmatrix} = \begin{pmatrix} \pm t + c \\ S(t)x + b(t) \end{pmatrix},$$

the transformation matrix $\left(\frac{\partial y'}{\partial x'}\right)$ has the form

$$\left(\frac{\partial y'}{\partial x'}\right) = \begin{pmatrix} \pm 1 & 0 & 0 & 0 \\ \dot{S}_{1j}(t)x^j + \dot{b}_1(t) & S_{11}(t) & S_{12}(t) & S_{13}(t) \\ \dot{S}_{2j}(t)x^j + \dot{b}_2(t) & S_{21}(t) & S_{22}(t) & S_{23}(t) \\ \dot{S}_{3j}(t)x^j + \dot{b}_3(t) & S_{31}(t) & S_{32}(t) & S_{33}(t) \end{pmatrix},$$

where j runs from 1 to 3. If we now calculate the transformation matrix $\left(\frac{\partial x'}{\partial y'}\right)$, we first have to keep in mind that from

$$s = \pm t + c \text{ and } y = S(t)x + b(t)$$

follows

$$t = \pm s \mp c \text{ and } x = S^{-1}(t)y - S^{-1}(t)b(t).$$

If we consider in addition that $S \in O(3)$ for every t, we get

$$x'(y') = \begin{pmatrix} \pm s \pm c \\ S^T(t)y - S^T(t)b(t) \end{pmatrix}.$$

In order to get the transformation matrix $\left(\frac{\partial x'}{\partial y'}\right)$ in dependence on t we also have to calculate:

$$\frac{dt}{ds} = \left(\frac{ds}{dt}\right)^{-1} = \pm 1$$

$$\frac{d}{ds}(S^T(t(s))) = \dot{S}^T(t) \cdot \frac{dt}{ds} = \pm \dot{S}^T(t)$$

$$\frac{d}{ds}(S^T(t(s))b(t(s))) = (\dot{S}^T(t)b(t) + S^T\dot{b}(t)) \cdot \frac{dt}{ds} = \pm(\dot{S}^T(t)b(t) + S^T\dot{b}(t))$$

$$\frac{d}{ds}(S^T(t(s))y - S^T(t(s))b(t(s))) = \pm\left(\dot{S}^T(t)y - \dot{S}^T(t)b(t) - S^T\dot{b}(t)\right).$$

Altogether, the transformation matrix $\left(\frac{\partial x'}{\partial y'}\right)$ has the form

$$\left(\frac{\partial x'}{\partial y'}\right) = \begin{pmatrix} \pm 1 & 0 & 0 & 0 \\ \pm\left(\dot{S}_{i1}(t)y^i - \dot{S}_{i1}(t)b^i(t) - S_{i1}(t)\dot{b}^i(t)\right) & S_{11}(t) & S_{21}(t) & S_{31}(t) \\ \pm\left(\dot{S}_{i2}(t)y^i - \dot{S}_{i2}(t)b^i(t) - S_{i2}(t)\dot{b}^i(t)\right) & S_{12}(t) & S_{22}(t) & S_{32}(t) \\ \pm\left(\dot{S}_{i3}(t)y^i - \dot{S}_{i3}(t)b^i(t) - S_{i3}(t)\dot{b}^i(t)\right) & S_{13}(t) & S_{23}(t) & S_{33}(t) \end{pmatrix}.$$

3.1. Transformation of coordinates

Again $i \in \{1,2,3\}$. Making use of all this it is now possible to study the behaviour of the Christoffel symbols with regard to our admitted transformations. In the following we will first consider only one of the Christoffel symbols contained in v. Then we know the transformation behaviour of v, since all components of v transform the same way.
Therefore, we now study the Christoffel symbol $\tilde{\Gamma}^3_{02}(y(x)) = \tilde{v}^1 \circ y(x)$. The other components of $\tilde{v}$ then arise from cyclical permutation. Note that only $\Gamma^k_{ij} \neq 0$, for $i = j = 0, k \neq 0$ or $i = 0, j \neq 0, k \neq 0$ or $i \neq 0, j = 0, k \neq 0$ respectively. If we use Greek letters in the following, the indices run from 1 to 3. Now we have:

$$\tilde{\Gamma}^3_{02} = \frac{\partial y^3}{\partial x^p} \frac{\partial x^m}{\partial y^0} \left(\frac{\partial^2 x^p}{\partial y^2 \partial y^l} \cdot \frac{\partial y^l}{\partial x^m} + \frac{\partial x^n}{\partial y^2} \Gamma^p_{mn} \right)$$

$$= \frac{\partial y^3}{\partial x^0} \frac{\partial x^m}{\partial y^0} \left(\underbrace{\frac{\partial^2 x^0}{\partial y^2 \partial y^l} \cdot \frac{\partial y^l}{\partial x^m}}_{=0} + \underbrace{\frac{\partial x^n}{\partial y^2} \Gamma^0_{mn}}_{=0} \right) + \frac{\partial y^3}{\partial x^\alpha} \frac{\partial x^m}{\partial y^0} \left(\frac{\partial^2 x^\alpha}{\partial y^2 \partial y^l} \cdot \frac{\partial y^l}{\partial x^m} + \frac{\partial x^n}{\partial y^2} \Gamma^\alpha_{mn} \right)$$

$$= \frac{\partial y^3}{\partial x^\alpha} \frac{\partial x^0}{\partial y^0} \left(\frac{\partial^2 x^\alpha}{\partial y^2 \partial y^l} \cdot \frac{\partial y^l}{\partial x^0} + \frac{\partial x^n}{\partial y^2} \Gamma^\alpha_{0n} \right) + \frac{\partial y^3}{\partial x^\alpha} \frac{\partial x^\beta}{\partial y^0} \left(\frac{\partial^2 x^\alpha}{\partial y^2 \partial y^l} \cdot \frac{\partial y^l}{\partial x^\beta} + \frac{\partial x^n}{\partial y^2} \underbrace{\Gamma^\alpha_{\beta n}}_{\Rightarrow n=0} \right)$$

$$= \frac{\partial y^3}{\partial x^\alpha} \frac{\partial x^0}{\partial y^0} \left(\frac{\partial^2 x^\alpha}{\partial y^2 \partial y^0} \cdot \frac{\partial y^0}{\partial x^0} + \underbrace{\frac{\partial^2 x^\alpha}{\partial y^2 \partial y^\gamma} \cdot \frac{\partial y^\gamma}{\partial x^0}}_{=0} + \underbrace{\frac{\partial x^0}{\partial y^2} \Gamma^\alpha_{00}}_{=0} + \frac{\partial x^\delta}{\partial y^2} \Gamma^\alpha_{0\delta} \right)$$

$$+ \frac{\partial y^3}{\partial x^\alpha} \frac{\partial x^\beta}{\partial y^0} \left(\frac{\partial^2 x^\alpha}{\partial y^2 \partial y^0} \cdot \underbrace{\frac{\partial y^0}{\partial x^\beta}}_{=0} + \underbrace{\frac{\partial^2 x^\alpha}{\partial y^2 \partial y^\gamma} \cdot \frac{\partial y^0}{\partial x^\beta}}_{=0} + \underbrace{\frac{\partial x^0}{\partial y^2} \Gamma^\alpha_{\beta 0}}_{=0} \right)$$

$$= \sum_{\alpha=1}^{3} (S_{3\alpha}(t) \cdot \dot{S}_{2\alpha}(t)) \pm \sum_{\alpha,\delta=1}^{3} S_{3\alpha}(t) \cdot S_{2\alpha}(t) \cdot \Gamma^\alpha_{0\delta}$$

$$= \sum_{\alpha=1}^{3} (S_{3\alpha}(t) \cdot \dot{S}_{2\alpha}(t)) \pm (S_{31}(t) \cdot S_{23}(t) \cdot \Gamma^1_{03} + S_{31}(t) \cdot S_{22}(t) \cdot \Gamma^1_{02} + S_{32}(t) \cdot S_{21}(t) \cdot \Gamma^2_{01}$$
$$+ S_{32}(t) \cdot S_{23}(t) \cdot \Gamma^2_{03} + S_{33}(t) \cdot S_{21}(t) \cdot \Gamma^3_{01} + S_{33}(t) \cdot S_{22}(t) \cdot \Gamma^3_{02})$$

$$= \sum_{\alpha=1}^{3} (S_{3\alpha}(t) \cdot \dot{S}_{2\alpha}(t)) \pm (\Gamma^3_{02}(S_{33}(t)S_{22}(t) - S_{32}(t)S_{23}(t))$$
$$+ \Gamma^1_{03}(S_{31}(t)S_{23}(t) - S_{33}(t)S_{21}(t)) + \Gamma^2_{01}(S_{32}(t)S_{21}(t) - S_{31}(t)S_{22}(t))).$$

For the other components of $\tilde{v}$ we get

$$\tilde{\Gamma}^1_{03} = \sum_{\alpha=1}^{3} (S_{1\alpha}(t) \cdot \dot{S}_{3\alpha}(t)) \pm (\Gamma^1_{03}(S_{11}(t)S_{33}(t) - S_{13}(t)S_{31}(t))$$
$$+ \Gamma^2_{01}(S_{12}(t)S_{31}(t) - S_{11}(t)S_{32}(t)) + \Gamma^3_{02}(S_{13}(t)S_{32}(t) - S_{12}(t)S_{33}(t)))$$

$$\tilde{\Gamma}^2_{01} = \sum_{\alpha=1}^{3}(S_{2\alpha}(t) \cdot \dot{S}_{1\alpha}(t)) \pm (\Gamma^2_{01}(S_{22}(t)S_{11}(t) - S_{21}(t)S_{12}(t))$$
$$+ \Gamma^3_{02}(S_{23}(t)S_{12}(t) - S_{22}(t)S_{13}(t)) + \Gamma^1_{03}(S_{21}(t)S_{13}(t) - S_{23}(t)S_{11}(t))).$$

After studying the transformation behaviour more closely, we are going to demonstrate a theorem which will make it easier to achieve our aim $v = 0$:

Theorem 3.1.1 *There is an adapted coordinate system so that $v = (\Gamma^3_{02}, \Gamma^1_{03}, \Gamma^2_{01})$ only depends on t if and only if there is an admitted coordinate transformation with*

$$\tilde{v} = (\tilde{\Gamma}^3_{02}, \tilde{\Gamma}^1_{03}, \tilde{\Gamma}^2_{01}) = 0.$$

Proof.

" $\Rightarrow$ ": Suppose, there is an adapted coordinate system so that $v = (\Gamma^3_{02}, \Gamma^1_{03}, \Gamma^2_{01})$ only depends on t. We now first consider the following:

Let $S \in O(3)$ be arbitrary. We now consider $O(3) =: G$ as Lie group and we try to demonstrate that the required transformation does exist.

Let $\mathcal{G} = so(3)$ be the Lie algebra belonging to G, so $\mathcal{G} = TG_E$. We first show that

$$TG_S = TG_E \cdot S.$$

Here we consider $O(3)$ as submanifold of $\text{Mat}_3(\mathbb{R})$ and therefore TG_E and TG_S as linear subspaces of $\text{Mat}_3(\mathbb{R})$.

" $\Rightarrow$ ": Let $B \in TG_S$. We choose a smooth curve $\beta : (-\epsilon, \epsilon) \to G$ which is appropriate for B. This means:
$$\beta(0) = S \text{ and } \dot{\beta}(0) = B.$$

Now consider the (smooth) curve $\alpha : (-\epsilon, \epsilon) \to G$ with
$$\alpha(u) = \beta(u) \cdot S^T.$$

Then
$$\alpha(0) = \beta(0) \cdot S^T = S \cdot S^T = E \text{ and so } \dot{\alpha}(t) \in TG_E.$$

But we also have $\dot{\alpha}(0) = B \cdot S^T$, altogether we get $B \cdot S^T \in TG_E$. So it immediately follows that
$$B = B \cdot E = B \cdot (S^T \cdot S) = (B \cdot S^T) \cdot S \in TG_E \cdot S.$$

" $\Leftarrow$ ": Now let $B \in TG_E \cdot S$ be arbitrary. Then there is an $A \in TG_E$ so that
$$B = A \cdot S.$$

3.1. Transformation of coordinates

Again let $\alpha : (-\epsilon, \epsilon) \to G$ be a smooth curve appropriate for A, so

$$\alpha(0) = E \text{ and } \dot\alpha(0) = A.$$

Then we have for B:

$$B = \dot\alpha(0) \cdot S.$$

We now choose the (smooth) curve $\beta : (-\epsilon, \epsilon) \to G$ with

$$\beta(u) = \alpha(u) \cdot S.$$

Then follows

$$\beta(0) = \alpha(0) \cdot S = E \cdot S = S \text{ and so } \dot\beta(0) \in TG_S.$$

On the other hand

$$B = \dot\alpha(0) \cdot S = \dot\beta(0) \in TG_S.$$

Now it is clear that $TG_S = TG_E \cdot S = \mathcal{G} \cdot S$. Then we have for $S(t) \in O(3)$

$$\dot S(t) \in TG_S = \mathcal{G} \cdot S.$$

So there is an $a(t) \in \mathcal{G}$ with $\dot S(t) = a(t) \cdot S$. But then we get

$$\dot S^T(t) = S^T \cdot a^T(t) = -S^T \cdot a(t),$$

since $a(t) \in \mathcal{G} = so(3)$, which are the skew symmetric matrices,

$$so(3) = \{a \in \mathrm{Mat}_3(\mathbb{R}) : a^T + a = 0\}.$$

But as $S \in O(3)$, we eventually get

$$S \cdot \dot S^T = -S \cdot S^T \cdot a(t) = -a(t) \in so(3).$$

It follows that $S \cdot \dot S^T$ is unambiguously determined by the components $(S \cdot \dot S^T)_{13}$, $(S \cdot \dot S^T)_{21}$ and $(S \cdot \dot S^T)_{32}$.

Let us now go back to our original problem. We want to show that - with the help of a special choice of $t \mapsto S(t)$ - we can achieve

$$0 = \sum_{\alpha=1}^{3} (S_{3\alpha}(t) \cdot \dot S_{2\alpha}(t)) \pm (\Gamma^3_{02}(S_{33}(t)S_{22}(t) - S_{32}(t)S_{23}(t))$$
$$+ \Gamma^1_{03}(S_{31}(t)S_{23}(t) - S_{33}(t)S_{21}(t)) + \Gamma^2_{01}(S_{32}(t)S_{21}(t) - S_{31}(t)S_{22}(t)))$$

3.1. Transformation of coordinates

$$0 = \sum_{\alpha=1}^{3}(S_{1\alpha}(t) \cdot \dot{S}_{3\alpha}(t)) \pm (\Gamma_{03}^{1}(S_{11}(t)S_{33}(t) - S_{13}(t)S_{31}(t))$$
$$+ \Gamma_{01}^{2}(S_{12}(t)S_{31}(t) - S_{11}(t)S_{32}(t)) + \Gamma_{02}^{3}(S_{13}(t)S_{32}(t) - S_{12}(t)S_{33}(t)))$$

$$0 = \sum_{\alpha=1}^{3}(S_{2\alpha}(t) \cdot \dot{S}_{1\alpha}(t)) \pm (\Gamma_{01}^{2}(S_{22}(t)S_{11}(t) - S_{21}(t)S_{12}(t))$$
$$+ \Gamma_{02}^{3}(S_{23}(t)S_{12}(t) - S_{22}(t)S_{13}(t)) + \Gamma_{03}^{1}(S_{21}(t)S_{13}(t) - S_{23}(t)S_{11}(t))).$$

If we write this in a different way we get

$$\sum_{\alpha=1}^{3}(S_{3\alpha}(t) \cdot \dot{S}_{2\alpha}(t)) = \mp (\Gamma_{02}^{3}(S_{33}(t)S_{22}(t) - S_{32}(t)S_{23}(t)) + \Gamma_{03}^{1}(S_{31}(t)S_{23}(t)$$
$$- S_{33}(t)S_{21}(t)) + \Gamma_{01}^{2}(S_{32}(t)S_{21}(t) - S_{31}(t)S_{22}(t)))$$

$$\sum_{\alpha=1}^{3}(S_{1\alpha}(t) \cdot \dot{S}_{3\alpha}(t)) = \mp (\Gamma_{03}^{1}(S_{11}(t)S_{33}(t) - S_{13}(t)S_{31}(t)) + \Gamma_{01}^{2}(S_{12}(t)S_{31}(t)$$
$$- S_{11}(t)S_{32}(t)) + \Gamma_{02}^{3}(S_{13}(t)S_{32}(t) - S_{12}(t)S_{33}(t)))$$

$$\sum_{\alpha=1}^{3}(S_{2\alpha}(t) \cdot \dot{S}_{1\alpha}(t)) = \mp (\Gamma_{01}^{2}(S_{22}(t)S_{11}(t) - S_{21}(t)S_{12}(t)) + \Gamma_{02}^{3}(S_{23}(t)S_{12}(t)$$
$$- S_{22}(t)S_{13}(t)) + \Gamma_{03}^{1}(S_{21}(t)S_{13}(t) - S_{23}(t)S_{11}(t))).$$

Now we can realize that these very components $(S \cdot \dot{S}^T)_{13}$, $(S \cdot \dot{S}^T)_{21}$ and $(S \cdot \dot{S}^T)_{32}$ are on the left hand side. The right hand side only depends on $S(t)$ and t. So we can put together the three equations to an ordinary differential equation on $so(3)$. For the right hand side we put $f(S,t)$ and so we get the equation

$$S \cdot \dot{S}^T = f(S,t),$$

which is equal to

$$\dot{S}^T = S^{-1} \cdot f(S,t) = S^T \cdot f(S,t),$$

and we finally get

$$\dot{S} = f(S,t)^T \cdot S =: g(S,t). \tag{3.1}$$

The existence of a solution for all $t \in \mathbb{R}$ follows from the compactness of $O(3)$. In fact, the differential equation (3.1) therefore is equivalent to the system

$$\begin{cases} \dot{v} = 1 \\ \dot{S} = g(S,v) \end{cases}$$

on $K \times \mathbb{R}$, $K := O(3)$ compact, with initial values

$$\begin{cases} v(0) = 0 \\ S(0) = E. \end{cases}$$

But this system has a unique solution $v(t) = t$ and $t \mapsto \widehat{S}(t)$ for all $t \in \mathbb{R}$ due to the usual theorems of the theory of ordinary differential equations.

As the transformation does not depend on $b : \mathbb{R} \to \mathbb{R}^3$ we can choose $b(t) = 0$ for all $t \in \mathbb{R}$. So we have found a solution $\widehat{S}(t)$ which exists for all $t \in \mathbb{R}$ and which fulfils

$$\dot{\widehat{S}}(t) = g(\widehat{S}(t), t).$$

If we now put the particular component of this solution in one of the three equations of the transformed Christoffel symbols we get, according to the construction of the solution,

$$\widetilde{v} = (\widetilde{\Gamma}^3_{02}, \widetilde{\Gamma}^1_{03}, \widetilde{\Gamma}^2_{01}) = 0,$$

which is what we wanted to show.

"$\Leftarrow$": Now let us suppose that there is an admitted coordinate transformation with $\widetilde{v} = 0$. Then v only depends on t since all terms which appear by any transformation depend on t only.

$\square$

Now we are also able to talk about the coordinate transformations which respect $v = 0$. The following theorem holds:

Theorem 3.1.2 *Let us suppose that there is an adapted coordinate system respect to which $v = 0$. Then the coordinate transformations of the form*

$$\tau : \mathbb{R} \times \mathbb{R}^3 \to \mathbb{R} \times \mathbb{R}^3, \quad (t, x) \mapsto (\pm t + c, Sx + b(t)),$$

where $c \in \mathbb{R}$, $b : \mathbb{R} \to \mathbb{R}^3$ and $S \in O(3)$, respect $v = 0$. These transformations are called Galiläi-transformations.

Proof. We already know that the components of v transform in the following way:

$$v_1 = \sum_{\alpha=1}^{3}(S_{3\alpha}(t) \cdot \dot{S}_{2\alpha}(t)) \pm (v_1(S_{33}(t)S_{22}(t) - S_{32}(t)S_{23}(t))$$
$$+ v_2(S_{31}(t)S_{23}(t) - S_{33}(t)S_{21}(t)) + v_3(S_{32}(t)S_{21}(t) - S_{31}(t)S_{22}(t)))$$
$$v_2 = \sum_{\alpha=1}^{3}(S_{1\alpha}(t) \cdot \dot{S}_{3\alpha}(t)) \pm (v_2(S_{11}(t)S_{33}(t) - S_{13}(t)S_{31}(t))$$
$$+ v_3(S_{12}(t)S_{31}(t) - S_{11}(t)S_{32}(t)) + v_1(S_{13}(t)S_{32}(t) - S_{12}(t)S_{33}(t)))$$

$$v_3 = \sum_{\alpha=1}^{3}(S_{2\alpha}(t) \cdot \dot{S}_{1\alpha}(t)) \pm (v_3(S_{22}(t)S_{11}(t) - S_{21}(t)S_{12}(t))$$
$$+ v_1(S_{23}(t)S_{12}(t) - S_{22}(t)S_{13}(t)) + v_2(S_{21}(t)S_{13}(t) - S_{23}(t)S_{11}(t))).$$

If we now suppose that there is an adapted coordinate system respect to which $v = 0$, this results in

$$v_1 = \sum_{\alpha=1}^{3}(S_{3\alpha}(t) \cdot \dot{S}_{2\alpha}(t)), \quad v_2 = \sum_{\alpha=1}^{3}(S_{1\alpha}(t) \cdot \dot{S}_{3\alpha}(t)), \quad v_3 = \sum_{\alpha=1}^{3}(S_{2\alpha}(t) \cdot \dot{S}_{1\alpha}(t)).$$

If we now want the coordinate transformation to respect $v = 0$, we have to consider

$$0 = \sum_{\alpha=1}^{3}(S_{3\alpha}(t) \cdot \dot{S}_{2\alpha}(t))$$
$$0 = \sum_{\alpha=1}^{3}(S_{1\alpha}(t) \cdot \dot{S}_{3\alpha}(t))$$
$$0 = \sum_{\alpha=1}^{3}(S_{2\alpha}(t) \cdot \dot{S}_{1\alpha}(t)).$$

But, of course, this is the case for $S \in O(3)$, which does not depend on t. □

Remark 3.1.3 Note that we can not say that these are the only coordinate transformations that respect $v = 0$. For instance, let $S : \mathbb{R} \to O(3)$ be of the form

$$S(t) = \begin{pmatrix} t & 0 & -1 \\ 0 & 1 & 0 \\ 1 & 0 & 0 \end{pmatrix}.$$

Then, $\det(S(t)) = 1$ holds for all $t \in \mathbb{R}$, thus $S(t) \in O(3)$, for all $t \in \mathbb{R}$. But the following also holds:

$$\sum_{\alpha=1}^{3}(S_{3\alpha}(t) \cdot \dot{S}_{2\alpha}(t)) = 0 \cdot 1 + 1 \cdot 0 + 0 \cdot 0 = 0$$
$$\sum_{\alpha=1}^{3}(S_{1\alpha}(t) \cdot \dot{S}_{3\alpha}(t)) = t \cdot 0 + 0 \cdot 0 + (-1) \cdot 0 = 0$$
$$\sum_{\alpha=1}^{3}(S_{2\alpha}(t) \cdot \dot{S}_{1\alpha}(t)) = 1 \cdot 0 + 0 \cdot 0 + 0 \cdot 0 = 0.$$

But, of course, S depends on t.

3.2 Conditions for the curvature tensor

In this section we want to assemble some conditions for the curvature tensor which are already known (see for example [10] and [42]).

Theorem 3.2.1 *Let (t, x^1, x^2, x^3) be an adapted coordinate system for M, $B := \frac{\partial}{\partial t}$ and $\nabla_{\partial_i} B = v_i^j \partial_j$. Then the following holds: v only depends on t if and only if $R(\partial_i, \partial_j)B = 0$, for all $1 \leq i, j \leq 3$.*

Proof.

"$\Rightarrow$" The condition that v only depends on t means that $D_i v_j^k = 0$, for all $1 \leq i, j, k \leq 3$. We now consider

$$R(\partial_i, \partial_j)B = \nabla_{\partial_i} \nabla_{\partial_j} B - \nabla_{\partial_j} \nabla_{\partial_i} B = \nabla_{\partial_i} v_j^k \partial_k - \nabla_{\partial_j} v_i^k \partial_k$$
$$= D_i v_j^k \partial_k + v_j^k \underbrace{\nabla_{\partial_i} \partial_k}_{=0} + D_j v_i^k \partial_k + v_i^k \underbrace{\nabla_{\partial_j} \partial_k}_{=0} = (D_i v_j^k - D_j v_i^k) \partial_k.$$

If v only depends on t this exactly means that

$$R(\partial_i, \partial_j)B = (D_i v_j^k - D_j v_i^k) \partial_k = 0.$$

"$\Leftarrow$" Now let $R(\partial_i, \partial_j)B = 0$, for all $1 \leq i, j \leq 3$. We know that for v

$$\text{rot}(v) = 0$$

holds, see (1.8). Therefore, we know that (locally) $v = \text{grad}(u)$ for a function u and we have

$$v_j^k = \epsilon_j^{kl} D_l u.$$

Due to the fact that $0 = R(\partial_i, \partial_j)B = (D_i v_j^k - D_j v_i^k) \partial_k$,

$$\epsilon_j^{kl} D_i D_l u - \epsilon_i^{kl} D_j D_l u = 0$$

holds, for all $1 \leq i, j, k, l \leq 3$. But this means that we get

$$D_i D_j u = 0$$

for all $1 \leq i, j \leq 3$, since, for instance,

$$0 = \epsilon_1^{23} D_2 D_3 u - \epsilon_2^{23} D_1 D_3 l u = D_2 D_3 u$$

and so on. But this means that v only depends on t.

$\square$

3.2. Conditions for the curvature tensor

Remark 3.2.2 If we remember the structure of (M, H) for $\lambda = 0$ (see chapter one), we already know that the leaves of the foliation are totally geodesic and flat, $R_p|_{H_p \times H_p \times H_p} = 0$. Therefore, the condition

$$R_p(\xi, \eta) B_p = 0,$$

for all $\xi, \eta \in H_p$ is equivalent to

$$R_p|_{H_p \times H_p} = 0,$$

for all $p \in M$. We therefore have

Corollary 3.2.3 v only depends on t if and only if $R_p|_{H_p \times H_p} = 0$, for all $p \in M$.

Trautman ([42]) gives us another equivalent formulation of the problem:

Theorem 3.2.4 Let (t, x^1, x^2, x^3) be an adapted coordinate system for M and $B := \frac{\partial}{\partial t}$. Then the following holds: $R(\partial_i, \partial_j)B = 0$, for all $1 \le i, j \le 3$, if and only if the bundle $H \to M$ is flat, $R^H = 0$.

Proof. We already know that $H \subseteq TM$ is totally geodesic, which implies that

$$R_p^H(\xi, \eta)\zeta = R_p(\xi, \eta)\zeta, \quad \text{for all } \xi, \eta \in TM_p \text{ and } \zeta \in H_p.$$

Furthermore, we already know that

$$R_p(\xi, \eta)\zeta = 0, \quad \text{for all } \xi, \eta, \zeta \in H_p,$$

since the leaves of H are flat. Thus, the condition that $H \to M$ is a flat bundle is equivalent to

$$R_p(B_p, \xi)\eta = 0, \quad \text{for all } \xi, \eta \in H_p, \tag{3.2}$$

since we get

$$R_p(\xi, \eta)\zeta = 0 \quad \text{for all } \xi, \eta \in TM_p \text{ and for all } \zeta \in H_p$$

due to the identities of R_p.

"$\Rightarrow$" Let $R(\partial_i, \partial_j)B = 0$ for all $1 \le i, j \le 3$ and let $\xi, \eta \in H_p$. Due to the first Bianchi identity

$$R_p(B_p, \xi)\eta = -R_p(\eta, B_p)\xi - \underbrace{R_p(\xi, \eta)B_p}_{=0} = R_p(B_p, \eta)\xi \tag{3.3}$$

holds. Now we consider the tri-linear map

$$\Phi_p : H_p \times H_p \times H_p \to \mathbb{R}, \quad (\xi, \eta, \zeta) \mapsto \tilde{h}_p(R_p(B_p, \xi)\eta, \zeta),$$

where $\tilde{h}$ is the bundle metric on H defined in the first chapter. As ∇^H is metric with respect to $\tilde{h}$, we know that $\tilde{h}_p(R_p(B_p, \xi)\eta, \zeta)$ is anti-symmetric in the last two arguments,

$$\tilde{h}_p(R_p(B_p, \xi)\eta, \zeta) = -\tilde{h}_p(R_p(B_p, \xi)\zeta, \eta),$$

but due to (3.3) also symmetric in ξ and η:

$$\tilde{h}_p(R_p(B_p,\xi)\eta,\zeta) = \tilde{h}_p(R_p(B_p,\eta)\xi,\zeta).$$

But this means that

$$\tilde{h}_p(R_p(B_p,\xi)\eta,\zeta) = \tilde{h}_p(R_p(B_p,\eta)\xi,\zeta) = -\tilde{h}_p(R_p(B_p,\eta)\zeta,\xi) = -\tilde{h}_p(R_p(B_p,\zeta)\eta,\xi)$$
$$= \tilde{h}_p(R_p(B_p,\zeta)\xi,\eta) = \tilde{h}_p(R_p(B_p,\xi)\zeta,\eta) = -\tilde{h}_p(R_p(B_p,\xi)\eta,\zeta).$$

Thus,

$$\tilde{h}_p(R_p(B_p,\xi)\eta,\zeta) = 0$$

and therefore

$$R_p(B_p,\xi)\eta = 0, \quad \text{for all } \xi,\eta \in H_p.$$

But, as already mentioned at the beginning of the proof, this is equivalent to the fact that $H \to M$ is a flat bundle.

"$\Leftarrow$" Now let $H \to M$ be a flat bundle, which means that (3.2) holds. But then (with the help of the first Bianchi identity)

$$R_p(\xi,\eta)B_p = -\underbrace{R_p(B_p,\xi)\eta}_{=0} - \underbrace{R_p(\eta,B_p)\xi}_{=0} = 0$$

holds, which means that $R(\partial_i,\partial_j)B = 0$, for all $1 \le i,j \le 3$.

$\square$

Corollary 3.2.5 *v only depends on t if and only if the bundle $H \to M$ is flat, $R^H = 0$.*

All in all, the following theorem holds:

Theorem 3.2.6 *Let (t,x^1,x^2,x^3) be an adapted coordinate system for M, $B := \frac{\partial}{\partial t}$ and $\nabla_{\partial_i} B = v_i^j \partial_j$. Then the following are equivalent:*

(i) v only depends on t.

(ii) $R(\partial_i,\partial_j)B = 0$, for all $1 \le i,j \le 3$.

(iii) $R_p|_{H_p \times H_p} = 0$, for all $p \in M$.

(iv) The bundle $H \to M$ is flat, $R^H = 0$.

Remark 3.2.7 The equivalent formulations mentioned above are of different qualities in the sense that some depend on a coordinate system and others do not. The assertions, for instance, "v only depends on t" or "$v = 0$" depend on the adapted coordinate system. Condition *(ii)*,

$R(\partial_i, \partial_j)B = 0$, for all $1 \leq i, j \leq 3$, does not depend on the choice of the coordinate system, since we can reformulate it as

$$R_p(\xi, \eta)B_p = 0, \quad \text{for all } \xi, \eta \in H_p \text{ and for all } p \in M,$$

but it does depend on the choice of B. Condition (iii) and (iv) do not depend on the adapted coordinate system or the choice of B. Condition (iv) is the most elegant equivalent as it just requires the flatness of a bundle.

3.3 Asymptotically flat spacetimes

In this section we want to discuss a further condition which ensures the existence of a genuine Newtonian limit. We first give a precise definition of asymptotically flat spacetimes and then adapt it to our case. It seems to be a natural assumption that isolated systems have a genuine Newtonian limit, since traditionally, Newtonian models only describe isolated systems. The mathematical equivalence to isolated systems is exactly the concept of asymptotically flat spacetimes.

Already in Ehlers' papers and Lottermoser's work it is mentioned that the right definition of asymptotically flat spacetimes might provide a condition for the existence of a genuine Newtonian limit. We here define the concept of asymptotically flat Ehlers spacetimes and show that it offers a condition for the existence of a genuine Newtonian limit. But we also show that our standard examples satisfy the definition of asymptotically flat Ehlers spacetimes, which guarantees that our definition is reasonable.

As a motivation, we first give a definition of asymptotically flat spacetimes used in General Relativity (see, for instance, [18]). Let therefore (M, g) be a spacetime.

Definition 3.3.1 Let $f : M \to \mathbb{R}$ be a global map, $c \in \mathbb{R}$ a regular value of f, so $df_p \neq 0$, for all $p \in M$, and $N \subseteq M$ the hypersurface defined by $N := \{p \in M : f(p) = c\}$.

(i) We first define the *co-normal* of (N, f) by $n := df|_N \in \Gamma(N, TM|_N)$. With regard to a coordinate system x on M in $p \in N$ we get for $n = (n_i)$, $n_i := \frac{\partial f}{\partial x^i}$ due to the fact that

$$n = df = \frac{\partial f}{\partial x^i} dx^i.$$

It induces a *normal vector* $\nu = (\nu^i)$ by $h(n, \_) = \langle \_, \nu \rangle$, where $\langle \_, \_ \rangle$ denotes the natural pairing and $h \in \Gamma(M, T^{(2,0)}M)$ the by g induced tensor field on $T^{(2,0)}M$. In coordinates we then get $\nu^i = h^{ij} n_j$.

(ii) A hypersurface N is called *spacelike*, if its normal vector is timelike.

3.3. Asymptotically flat spacetimes

Definition 3.3.2 The *standard Lorentzian metric* $\eta = (\eta_{ij})$ on $\mathbb{R} \times \mathbb{R}^3$ is given by

$$\eta_{ij} = \begin{cases} -1 & \text{for } i = j = 0 \\ 1 & \text{for } i = j \geq 1 \\ 0 & \text{otherwise.} \end{cases}$$

Definition 3.3.3 (i) We call $\mathbb{B}_R(0) := \{p \in \mathbb{R}^3 : ||p|| \leq R\}$, with $R > 0$ the three-dimensional *ball around zero of radius R*.

(ii) The four-dimensional *cylinder around zero of radius R* is given by $Z_R(0) := \mathbb{R} \times \mathbb{B}_R(0)$.

Definition 3.3.4 A spacetime (M, g) is called *asymptotically flat at spacelike infinity* with regard to a map $x : M \setminus A \to \mathbb{R}^4 \setminus Z_R(0)$ (short: *asymptotically flat*), where $A \subseteq M$ is a closed subset and:

(i) $N_t := \{p \in M \setminus A : x^0(p) = t\} \subseteq M \setminus A$ is a spacelike hypersurface in $M \setminus A$ for all $t \in \mathbb{R}$, where $\text{im}(x^0) = \mathbb{R}$ and all $t \in \mathbb{R}$ are regular values of x^0. We call N_t a *spaceleaf*.

(ii) For every spaceleaf we have the fading out conditions
$(||x|| : U \subseteq \mathbb{R}^3 \to (0, \infty), \ ||x|| := \sqrt{\sum_{i=1}^{3}(x^i)^2})$:

a) $g_{ij} - \eta_{ij} = \mathcal{O}(||x||^{-1})$, $0 \leq i, j \leq 3$,

b) $\partial_k g_{ij} = \mathcal{O}(||x||^{-2})$, $0 \leq i, j, k \leq 3$,

c) $\partial_l \partial_k g_{ij} = \mathcal{O}(||x||^{-3})$, $0 \leq i, j, k, l \leq 3$,

uniformly on compact x^0- intervals.

As mentioned before, the definition stems from ([18]).

Remark 3.3.5 In this definition, $g_{ij} - \eta_{ij} = \mathcal{O}(||x||^{-1})$, $0 \leq i, j \leq 3$, means: for all $t \in \mathbb{R}$ and for all $0 \leq i, j \leq 3$ there is a $c_{ij}(t) > 0$ and an $r_0 > 0$ so that

$$|g_{ij}(t, \overline{x}) - \eta_{ij}| \leq \frac{c_{ij}(t)}{||x||}, \quad \forall \overline{x} \text{ with } |\overline{x}| \geq r_0.$$

"Uniformly on compact x^0- intervals I" means that the time-dependent maps $c_{ij}(t)$ are limited on I by a constant $C > 0$ (possibly depending on I), so

$$c_{ij}(t) < C, \quad \forall 0 \leq i, j \leq 3, \quad \forall t \in I.$$

The same holds for the higher derivations respectively.

Now we have to adapt the concept of asymptotically flat spacetimes to our case. We define when a family of Ehlers spacetimes which has a quasi-Newtonian limit is called asymptotically flat. Thus, we first have to define the standard Lorentzian metric in dependence on λ:

3.3. Asymptotically flat spacetimes

Definition 3.3.6 The *standard Lorentzian metric* $\eta(\lambda) = (\eta_{ij}(\lambda))$ on $\mathbb{R} \times \mathbb{R}^3$ in dependence on λ is given by

$$\eta_{ij}(\lambda) = \begin{cases} 1 & \text{for } i = j = 0 \\ -\lambda & \text{for } i = j \geq 1 \\ 0 & \text{otherwise.} \end{cases}$$

Remark 3.3.7 a) In the following definition we will see that we have to compare Ehlers spacetimes with the standard Lorentzian metric and its derivations. If we bear in mind that many solutions of Einstein's field equations (and therefore also their extensions) use spherical coordinates it is sometimes easier to consider the Lorentzian metric in spherical coordinates. Then the standard Lorentzian metric in dependence on λ has the form:

$$\eta_{ij}(\lambda) = \text{diag}(1, -\lambda, -\lambda r^2, -\lambda r^2 \sin^2 \vartheta).$$

b) If we have a look at the index and signature of $\eta_{ij}(\lambda)$, we see that it is not a Lorentzian metric in the original sense (it has index 3). But we have to remember that in case of $\lambda > 0$, h has the structure of a Lorentzian metric and g does not have the right index. But as for $\lambda > 0$ the two metrics are non-degenerate and almost inverse it makes no difference whether we discuss the following for g or h.

c) Now we can adapt the definition of asymptotically flat spacetimes to our case. We have to remember that a family of Ehlers spacetimes $M_\lambda = (M, g(\lambda), h(\lambda), \nabla(\lambda), T(\lambda), \lambda)$, $\lambda \in (0, 1]$, has a quasi-Newtonian limit for $\lambda = 0$, if all the fields have a pointwise limit, the convergence is of right order and if the limit satisfies the frame theory for $\lambda = 0$.

Definition 3.3.8 Let $M_\lambda = (M, g(\lambda), h(\lambda), \nabla(\lambda), T(\lambda), \lambda)$, $\lambda \in (0, 1]$ be a family of Ehlers spacetimes which has a quasi-Newtonian limit. We call this family *asymptotically flat at spacelike infinity*, if there is a global coordinate system $(t, x) : M \to I \times \mathbb{R}^3$ so that

(i) for every spaceleaf $N_{t_0} := \{p \in M : t(p) = t_0\} \subseteq M$ the fading out conditions

   a) $g_{ij}(\lambda, t, x) - \eta_{ij}(\lambda, t, x) = \mathcal{O}(||x||^{-1})$, $0 \leq i, j \leq 3$,

   b) $\Gamma^k_{ij}(\lambda, t, x) = \mathcal{O}(||x||^{-2})$, $0 \leq i, j, k \leq 3$,

   c) $R^l_{ijk}(\lambda, t, x) = \mathcal{O}(||x||^{-3})$, $0 \leq i, j, k, l \leq 3$

   hold, uniformly on compact $t-$ intervals;

(ii) for $\lim_{\lambda \to 0} M_\lambda =: M_0$ these global coordinates turn into special Newtonian ones;

(iii) the limits for $\lambda \to 0$ and $||x|| \to \infty$ can be exchanged.

Remark 3.3.9 a) In the first chapter we have already seen that in case of $\lambda = 0$ we can detect special Newtonian coordinates. With regard to these coordinates we calculated the quasi-Newtonian equations and v. At the beginning of this chapter we discussed which coordinate transformations are permitted in case of $\lambda = 0$.

b) This definition of asymptotically flat Ehlers spacetimes includes the fact that the respective spacetimes for λ fixed are asymptotically flat in the former sense. Furthermore, it demands compatibility with the limit process and thus with the transition from $\lambda > 0$ to $\lambda = 0$. This is not surprising since the structures of the manifold have to turn into the Newtonian structure.

c) Our standard examples Minkowski (of course), Schwarzschild and Kerr satisfy this definition. This will be discussed more closely in (3.3.11).

Theorem 3.3.10 *Let $M_\lambda = (M, g(\lambda), h(\lambda), \nabla(\lambda), T(\lambda), \lambda)$, $\lambda \in (0, 1]$ be a family of Ehlers spacetimes which has a quasi-Newtonian limit. Let the family of Ehlers spacetimes be asymptotically flat at spacelike infinity. Then the limit is a genuine Newtonian one.*

Proof. As the family of Ehlers spacetimes is asymptotically flat, we know from definition (3.3.8) that there are global coordinates $(t, x) : M \to I \times \mathbb{R}^3$. Therefore, the manifold M is already simply connected and geodesically complete in space direction. Thus, we just have to show that $v(x, t) = 0$. We therefore consider the Christoffel symbols $\Gamma^3_{02}(\lambda, t, x)$, $\Gamma^1_{03}(\lambda, t, x)$ and $\Gamma^2_{01}(\lambda, t, x)$ for $\lambda \in (0, 1]$ in the special coordinates given by the definition of asymptotically flat spacetimes. This definition then provides that for $\lambda \in (0, 1]$

$$\Gamma^3_{02}(\lambda, t, x) \to 0, \ \Gamma^1_{03}(\lambda, t, x) \to 0 \text{ and } \Gamma^2_{01}(\lambda, t, x) \to 0,$$

for $||x|| \to \infty$.

In addition we can exchange the limits for $\lambda \to 0$ and $||x|| \to \infty$ due to definition (3.3.8). Then

$$\Gamma^3_{02}(0, t, x) \to, \ \Gamma^1_{03}(0, t, x) \to 0 \text{ and } \Gamma^2_{01}(0, t, x) \to 0, \tag{3.4}$$

for $||x|| \to \infty$.

Furthermore, we have to consider that $\Delta v = 0$ for t fixed, since

$$\Delta v = \operatorname{grad}(\operatorname{div} v) - \operatorname{rot}(\operatorname{rot} v).$$

As shown in the first chapter, $\operatorname{div}(v) = 0$ and $\operatorname{rot}(v) = 0$ hold for v in special Newtonian coordinates (see equations (1.12) and (1.8)). As the definition of asymptotically flat spacetimes requires that the global coordinates turn into special Newtonian ones for $\lambda \to 0$, this condition is satisfied and we actually get

$$\Delta v = \operatorname{grad}(\operatorname{div} v) - \operatorname{rot}(\operatorname{rot} v) = 0.$$

If we have a look at the definition of Δv,

$$\Delta v = \begin{pmatrix} \frac{\partial^2 v^1}{(\partial x^1)^2} + \frac{\partial^2 v^1}{(\partial x^2)^2} + \frac{\partial^2 v^1}{(\partial x^3)^2} \\ \frac{\partial^2 v^2}{(\partial x^1)^2} + \frac{\partial^2 v^2}{(\partial x^2)^2} + \frac{\partial^2 v^2}{(\partial x^3)^2} \\ \frac{\partial^2 v^3}{(\partial x^1)^2} + \frac{\partial^2 v^3}{(\partial x^2)^2} + \frac{\partial^2 v^3}{(\partial x^3)^2} \end{pmatrix} = \begin{pmatrix} \Delta v^1 \\ \Delta v^2 \\ \Delta v^3 \end{pmatrix},$$

3.3. Asymptotically flat spacetimes

$\Delta v = 0$ implies $\Delta v^i = 0$ for $i \in \{1, 2, 3\}$.
Finally, we know with the help of Hopf's maximum principle that $v^i(\_, t) : \mathbb{R}^3 \to \mathbb{R}$ is constant, thus, due to (3.4), $v^i(x) = 0$ for every (x, t) and $i \in \{1, 2, 3\}$. This shows the claim. □

Example 3.3.11 Now we want to show that the definition of asymptotically flat families is reasonable. This is the case, if the standard examples fulfil the conditions of the definition. We do not have to consider Minkowski spacetime as it obviously meets the conditions. Therefore, we discuss the extended Schwarzschild and Kerr case.

We consider the Minkowski spacetime in spherical coordinates. Then $g^M(\lambda, r, \vartheta, \varphi)$ has the form
$$(g^M(\lambda, r, \vartheta, \varphi)_{ij}) = \operatorname{diag}(1, -\lambda, -\lambda r^2, -\lambda r^2 \sin^2 \vartheta),$$
where $r > 0$, $\vartheta \in (0, \pi)$, $\varphi \in (0, 2\pi)$. The Christoffel symbols in spherical coordinates which do not vanish have the form

$$\Gamma^1_{22}(\lambda, r, \vartheta, \varphi) = -r$$
$$\Gamma^2_{12}(\lambda, r, \vartheta, \varphi) = \frac{1}{r}$$
$$\Gamma^3_{13}(\lambda, r, \vartheta, \varphi) = \frac{1}{r}$$
$$\Gamma^1_{33}(\lambda, r, \vartheta, \varphi) = -r \sin^2 \vartheta$$
$$\Gamma^2_{33}(\lambda, r, \vartheta, \varphi) = -\sin \vartheta \cos \vartheta$$
$$\Gamma^3_{23}(\lambda, r, \vartheta, \varphi) = \frac{\cos \vartheta}{\sin \vartheta}.$$

a) Schwarzschild spacetime (see (A.1) and (2.3.2)):

First we have a look at the metric components. The only differences which do not vanish are the following:

$$|g^S_{00}(\lambda) - g^M_{00}| = \left|1 - \frac{2\lambda}{r} - 1\right| = \frac{2\lambda}{r}$$
$$|g^S_{11}(\lambda) - g^M_{11}| = \left|\frac{-\lambda r}{r - 2\lambda} + \lambda\right| = \left|\frac{-\lambda r + \lambda r - 2\lambda^2}{r - 2\lambda}\right| = \frac{2\lambda^2}{r - 2\lambda},$$

which are of order $\left(\frac{1}{r}\right)$, of course. For the Christoffel symbols we have

$$|{}^S\Gamma^1_{00}(\lambda) - {}^M\Gamma^1_{00}(\lambda)| = \left|\frac{r - 2\lambda}{r^3}\right| = \left|\frac{1}{r^2} - \frac{2\lambda}{r^3}\right|$$
$$|{}^S\Gamma^0_{10}(\lambda) - {}^M\Gamma^0_{10}(\lambda)| = \left|\frac{\lambda}{r^2 - 2\lambda r}\right|$$
$$|{}^S\Gamma^1_{11}(\lambda) - {}^M\Gamma^1_{11}(\lambda)| = \left|\frac{\lambda}{r^2 - 2\lambda r}\right|$$
$$|{}^S\Gamma^1_{22}(\lambda) - {}^M\Gamma^1_{22}(\lambda)| = |2\lambda - r + r| = 2\lambda$$
$$|{}^S\Gamma^2_{12}(\lambda) - {}^M\Gamma^2_{12}(\lambda)| = \left|\frac{1}{r} - \frac{1}{r}\right| = 0$$

3.3. Asymptotically flat spacetimes

$$|{}^S\Gamma^1_{33}(\lambda) - {}^M\Gamma^1_{33}(\lambda)| = |2\lambda \sin^2\vartheta - r\sin^2\vartheta + r\sin^2\vartheta| = 2\lambda \sin^2\vartheta$$

$$|{}^S\Gamma^3_{13}(\lambda) - {}^M\Gamma^3_{13}(\lambda)| = \left|\frac{1}{r} - \frac{1}{r}\right| = 0$$

$$|{}^S\Gamma^2_{33}(\lambda) - {}^M\Gamma^2_{33}(\lambda)| = |-\sin\vartheta\cos\vartheta + \sin\vartheta\cos\vartheta| = 0$$

$$|{}^S\Gamma^1_{00}(\lambda) - {}^M\Gamma^1_{00}(\lambda)| = \left|\frac{\cos\vartheta}{\sin\vartheta} - \frac{\cos\vartheta}{\sin\vartheta}\right| = 0.$$

If we have a look at this list we see that the differences are zero, of order $\left(\frac{1}{r^2}\right)$ or constant. In case of $\Gamma^1_{22}(\lambda)$ and $\Gamma^1_{33}(\lambda)$ the condition $(i)b$ of definition (3.3.8) is not satisfied. But this does not depend on the extension of the Schwarzschild spacetime as for the standard Schwarzschild case the following holds:

$$|{}^S\Gamma^1_{22} - {}^M\Gamma^1_{22}| = |2m - r + r| = 2m$$

$$|{}^S\Gamma^1_{33} - {}^M\Gamma^1_{33}| = |2m\sin^2\vartheta - r\sin^2\vartheta + r\sin^2\vartheta| = 2m\sin^2\vartheta.$$

Thus, the standard Schwarzschild spacetime in spherical coordinates does not satisfy the definition of being asymptotically flat in this point, which means that spherical coordinates are not the suitable ones. Therefore, we have to transform the metric and the Christoffel symbols into cartesian coordinates.

Let $u = (t, r, \vartheta, \varphi)$ be spherical and $x = (t, x^1, x^2, x^3)$ cartesian coordinates. Then the transformation matrices have the form

$$\left(\frac{\partial x}{\partial u}\right) = \begin{pmatrix} 1 & 0 & 0 & 0 \\ 0 & \cos\vartheta\sin\varphi & -r\sin\vartheta\sin\varphi & r\cos\vartheta\cos\varphi \\ 0 & \sin\vartheta\sin\varphi & r\cos\vartheta\sin\varphi & r\sin\vartheta\cos\varphi \\ 0 & \cos\varphi & 0 & -r\sin\varphi \end{pmatrix},$$

and

$$\left(\frac{\partial u}{\partial x}\right) = \begin{pmatrix} 1 & 0 & 0 & 0 \\ 0 & \cos\vartheta\sin\varphi & \sin\vartheta\sin\varphi & \cos\varphi \\ 0 & -\frac{\sin\vartheta}{r\sin\varphi} & \frac{\cos\vartheta}{r\sin\varphi} & 0 \\ 0 & \frac{\cos\vartheta\cos\varphi}{r} & \frac{\sin\vartheta\cos\varphi}{r} & -\frac{\sin\varphi}{r} \end{pmatrix}.$$

First we have a look at the transformation behaviour of the metric:

$$g_{ij}(x(u)) = \frac{\partial u^\alpha}{\partial x^i}(x) g_{\alpha\beta}(u(x)) \frac{\partial u^\beta}{\partial x^j}(x).$$

Since the matrix $\left(\frac{\partial u}{\partial x}\right)$ only has components which are either constant with regard to r or of order $\left(\frac{1}{r}\right)$ the transformation just improves the behaviour for $r \to \infty$. As the components of the metric already have the right vanishing behaviour, we do not have to care about them.

Thus, let us have a look at the Christoffel symbols. We already know their transformation

behaviour:
$$\widetilde{\Gamma}^k_{ij}(x(u)) = \left(\frac{\partial x^k}{\partial u^p} \cdot \frac{\partial u^m}{\partial x^i}(x)\left(\frac{\partial^2 u^p}{\partial x^j \partial x^l}(x) \cdot \frac{\partial x^l}{\partial u^m} + \frac{\partial u^n}{\partial x^j}(x) \cdot \Gamma^p_{mn}\right)\right)(u).$$

As we only want to understand the behaviour of the differences ${}^S\Gamma^k_{ij} - {}^M\Gamma^k_{ij}$, we just have to consider the last part of this transformation formula, since the first part vanishes. This means, we have to look at the following:

$$ {}^S\widetilde{\Gamma}^k_{ij} - {}^M\widetilde{\Gamma}^k_{ij} = \frac{\partial x^k}{\partial u^p} \cdot \left(\frac{\partial u^m}{\partial x^i} \frac{\partial u^n}{\partial x^j}\right) \cdot ({}^S\Gamma^p_{mn} - {}^M\Gamma^p_{mn}).$$

We first put forward an argument which shows that the vanishing behaviour of the Christoffel symbols does not get worse. We then show that the two Christoffel symbols which do not have the right vanishing behaviour in spherical coordinates have the right behaviour after the transformation.

For all the differences ${}^S\Gamma^p_{mn} - {}^M\Gamma^p_{mn}$ in spherical coordinates which do not vanish, p equals either 0 or 1. We therefore multiply the differences with components of $\left(\frac{\partial u}{\partial x}\right)$ (which are either constant with regard to r or of order $\left(\frac{1}{r}\right)$) and with components of the first two columns of $\left(\frac{\partial x}{\partial u}\right)$, which do not depend on r. Thus, the vanishing behaviour does not get worse.

Now we have a look at the two differences which do not have the right behaviour. After transformation they are of the following form:

$$({}^S\Gamma^1_{22} - {}^M\Gamma^1_{22}) = \left(\frac{\partial u^1}{\partial x^2}\right)^2 \cdot \frac{\partial x^1}{\partial u^1} \cdot \frac{\lambda}{r^2 - 2\lambda r} + \left(\frac{\partial u^2}{\partial x^2}\right)^2 \cdot \frac{\partial x^1}{\partial u^1} \cdot 2\lambda$$
$$+ \left(\frac{\partial u^3}{\partial x^2}\right)^2 \cdot \frac{\partial x^1}{\partial u^1} \cdot 2\lambda \sin^2 \vartheta$$
$$= \frac{\lambda \cdot \sin^2 \vartheta \sin^3 \varphi \cos \vartheta}{r^2 - 2\lambda r} + \frac{2\lambda \cdot \cos^3 \vartheta}{r^2 \sin \varphi}$$
$$+ \frac{2\lambda \cdot \sin^4 \vartheta \cos \vartheta \sin \varphi \cos^2 \varphi}{r^2} = \mathcal{O}\left(\frac{1}{r^2}\right)$$

$$({}^S\Gamma^1_{33} - {}^M\Gamma^1_{33}) = \left(\frac{\partial u^1}{\partial x^3}\right)^2 \cdot \frac{\partial x^1}{\partial u^1} \cdot \frac{\lambda}{r^2 - 2\lambda r} + \left(\frac{\partial u^2}{\partial x^3}\right)^2 \cdot \frac{\partial x^1}{\partial u^1} \cdot 2\lambda$$
$$+ \left(\frac{\partial u^3}{\partial x^3}\right)^2 \cdot \frac{\partial x^1}{\partial u^1} \cdot 2\lambda \sin^2 \vartheta$$
$$= \frac{\lambda \cdot \cos \vartheta \sin \varphi \cos^2 \varphi}{r^2 - 2\lambda r} + 0 + \frac{2\lambda \cdot \sin^2 \vartheta \cos \vartheta \sin^3 \varphi}{r^2} = \mathcal{O}\left(\frac{1}{r^2}\right).$$

This shows that the Christoffel symbols have the right vanishing behaviour and therefore meet the conditions of the definition of asymptotically flat families of Ehlers spacetimes. Now we have to discuss the vanishing behaviour of the components of the curvature tensor. As the curvature tensor of the Minkowski spacetime vanishes for all coordinates (remember that the components vanish for cartesian coordinates and, as it is a tensor,

3.3. Asymptotically flat spacetimes

also for all the other coordinates), we only have to consider the vanishing behaviour of the components of the curvature tensor of Schwarzschild spacetime. The ones which do not vanish immediately are (see (A.1.5)):

$$|R^0_{101}| = \left|\frac{-2\lambda r}{(r^2 - 2\lambda r)^2}\right| = \mathcal{O}\left(\frac{1}{r^3}\right)$$

$$|R^0_{202}| = \left|\frac{\lambda}{r}\right| = \mathcal{O}\left(\frac{1}{r}\right)$$

$$|R^0_{303}| = \left|-\frac{\lambda \sin^2 \vartheta}{r}\right| = \mathcal{O}\left(\frac{1}{r}\right)$$

$$|R^1_{010}| = \left|\frac{4\lambda - 2r}{r^4}\right| = \mathcal{O}\left(\frac{1}{r^3}\right)$$

$$|R^2_{020}| = \left|\frac{r - 2\lambda}{r^4}\right| = \mathcal{O}\left(\frac{1}{r^3}\right)$$

$$|R^3_{030}| = \left|\frac{r - 2\lambda}{r^4}\right| = \mathcal{O}\left(\frac{1}{r^3}\right)$$

$$|R^1_{212}| = \left|\frac{-\lambda}{r}\right| = \mathcal{O}\left(\frac{1}{r}\right)$$

$$|R^2_{121}| = \left|\frac{\lambda}{r^2(2\lambda - r)}\right| = \mathcal{O}\left(\frac{1}{r^3}\right)$$

$$|R^1_{313}| = \left|\frac{-\lambda \sin^2 \vartheta}{r}\right| = \mathcal{O}\left(\frac{1}{r}\right)$$

$$|R^3_{131}| = \left|\frac{\lambda}{r^2(2\lambda - r)}\right| = \mathcal{O}\left(\frac{1}{r^3}\right)$$

$$|R^2_{323}| = \left|\sin^2 \vartheta \cdot \frac{2\lambda}{r}\right| = \mathcal{O}\left(\frac{1}{r}\right)$$

$$|R^3_{232}| = \left|\frac{2\lambda}{r}\right| = \mathcal{O}\left(\frac{1}{r}\right).$$

Now we again have to consider the transformation behaviour. For the curvature tensor $\widetilde{R}^l_{ijk} = \frac{\partial u^\alpha}{\partial x^i}\frac{\partial u^\beta}{\partial x^j}\frac{\partial u^\gamma}{\partial x^k} R^\delta_{\alpha\beta\gamma} \frac{\partial x^l}{\partial u^\delta}$ holds. As the $\widetilde{R}^l_{ijk}$ are sums of the transformed $R^\delta_{\alpha\beta\gamma}$, we just have a look at the vanishing behaviour of the latter. Therefore, if δ is 1 or 0, the behaviour does not deteriorate. So we do not have to care about those components with $\delta = 0$ or $\delta = 1$ which already have the right behaviour. For the other ones we have to count how many of the indices α, β or γ equal 2 or 3. Then the transformed component either equals zero or the vanishing behaviour improves (for every index which equals 2 or 3 we get an $\mathcal{O}\left(\frac{1}{r}\right)$). Then the following holds (if the components do not vanish):

$$|R^0_{101}| = \mathcal{O}\left(\frac{1}{r^3}\right)$$

$$|R^0_{202}| = \mathcal{O}\left(\frac{1}{r^2}\right) \cdot \mathcal{O}\left(\frac{1}{r^2}\right) = \mathcal{O}\left(\frac{1}{r^4}\right)$$

$$|R^0_{303}| = \mathcal{O}\left(\frac{1}{r^2}\right) \cdot \mathcal{O}\left(\frac{1}{r^2}\right) = \mathcal{O}\left(\frac{1}{r^4}\right)$$

$$|R^1_{010}| = \mathcal{O}\left(\frac{1}{r^3}\right)$$

$$|R^2_{020}| = \mathcal{O}\left(\frac{1}{r}\right) \cdot \mathcal{O}\left(\frac{1}{r^3}\right) \cdot \mathcal{O}(r) = \mathcal{O}\left(\frac{1}{r^3}\right)$$

$$|R^3_{030}| = \mathcal{O}\left(\frac{1}{r}\right) \cdot \mathcal{O}\left(\frac{1}{r^3}\right) \cdot \mathcal{O}(r) = \mathcal{O}\left(\frac{1}{r^3}\right)$$

$$|R^1_{212}| = \mathcal{O}\left(\frac{1}{r^2}\right) \cdot \mathcal{O}\left(\frac{1}{r^2}\right) = \mathcal{O}\left(\frac{1}{r^4}\right)$$

$$|R^2_{121}| = \mathcal{O}\left(\frac{1}{r}\right) \cdot \mathcal{O}\left(\frac{1}{r^3}\right) \cdot \mathcal{O}(r) = \mathcal{O}\left(\frac{1}{r^3}\right)$$

$$|R^1_{313}| = \mathcal{O}\left(\frac{1}{r^2}\right) \cdot \mathcal{O}\left(\frac{1}{r^2}\right) = \mathcal{O}\left(\frac{1}{r^4}\right)$$

$$|R^3_{131}| = \mathcal{O}\left(\frac{1}{r}\right) \cdot \mathcal{O}\left(\frac{1}{r^3}\right) \cdot \mathcal{O}(r) = \mathcal{O}\left(\frac{1}{r^3}\right)$$

$$|R^2_{323}| = \mathcal{O}\left(\frac{1}{r^3}\right) \cdot \mathcal{O}\left(\frac{1}{r^3}\right) \cdot \mathcal{O}(r) = \mathcal{O}\left(\frac{1}{r^5}\right)$$

$$|R^3_{232}| = \mathcal{O}\left(\frac{1}{r^3}\right) \cdot \mathcal{O}\left(\frac{1}{r^3}\right) \cdot \mathcal{O}(r) = \mathcal{O}\left(\frac{1}{r^5}\right).$$

This means that the components of the curvature tensor have the required vanishing behaviour and the extended Schwarzschild spacetime here also satisfies the definition of asymptotically flat spacetimes.

If we have a look at the Christoffel symbols for $\lambda \in (0, 1]$ which do not vanish,

$$\Gamma^1_{11} = \frac{\lambda}{2\lambda r - r^2}$$

$$\Gamma^1_{00} = \frac{r - 2\lambda}{r^3}$$

$$\Gamma^1_{22} = 2\lambda - r$$

$$\Gamma^1_{33} = (2\lambda - r)\sin^2\vartheta$$

$$\Gamma^2_{33} = -\sin\vartheta\cos\vartheta$$

$$\Gamma^0_{10} = \frac{\lambda}{r^2 - 2\lambda r}$$

$$\Gamma^2_{12} = \frac{1}{r}$$

$$\Gamma^3_{13} = \frac{1}{r}$$

$$\Gamma^3_{23} = \frac{\cos\vartheta}{\sin\vartheta},$$

we see that there is no problem if we want to exchange the limits for $\lambda \to 0$ and $r \to \infty$. Thus, all in all, we can say that the extended Schwarzschild spacetime satisfies the definition of asymptotically flat spacetimes.

3.3. Asymptotically flat spacetimes

b) Kerr spacetime (see (A.2) and (2.3.3)):
Again we first have a look at the metric components. We have:

$$|g_{00}^K(\lambda) - g_{00}^M(\lambda)| = \left|1 - \frac{2\lambda r}{r^2 + a^2\cos^2\vartheta} - 1\right| = \frac{2\lambda r}{r^2 + a^2\cos^2\vartheta}$$

$$|g_{03}^K(\lambda) - g_{03}^M(\lambda)| = \left|\frac{2\lambda^{\frac{3}{2}} r a \sin^2\vartheta}{r^2 + a^2\cos^2\vartheta}\right| = \frac{2\lambda^{\frac{3}{2}} r a \sin^2\vartheta}{r^2 + a^2\cos^2\vartheta}$$

$$|g_{11}^K(\lambda) - g_{11}^M(\lambda)| = \left|-\lambda\frac{r^2 + a^2\cos^2\vartheta}{r^2 - 2\lambda r + a^2} + \lambda\right|$$

$$= \left|\lambda\frac{r^2 - 2\lambda r + a^2 - r^2 - a^2\cos^2\vartheta}{r^2 - 2\lambda r + a^2}\right| = \left|\lambda\frac{-2\lambda r + a^2\sin^2\vartheta}{r^2 - 2\lambda r + a^2}\right|$$

$$|g_{22}^K(\lambda) - g_{22}^M(\lambda)| = |-\lambda(r^2 + a^2\cos^2\vartheta) + \lambda r^2| = \lambda a^2\cos^2\vartheta$$

$$|g_{33}^K(\lambda) - g_{33}^M(\lambda)| = \left|-\lambda\sin^2\vartheta\left(r^2 + a^2 + \frac{2\lambda r a^2\sin^2\vartheta}{r^2 + a^2\cos^2\vartheta}\right) + \lambda r^2\sin^2\vartheta\right|$$

$$= \left|-\lambda a^2\sin^2\vartheta - \lambda\sin^2\vartheta\frac{2\lambda r a^2\sin^2\vartheta}{r^2 + a^2\cos^2\vartheta}\right|.$$

Here we also have the problem that the two last differences do not satisfy the first condition of definition (3.3.8). Again we have to discuss the transformation behaviour. As we already know the transformation (which is the same as in case of Schwarzschild) only improves the behaviour. For the critical components of the metric we get:

$$g_{22}^K - g_{22}^M = \left(\frac{\partial u^1}{\partial x^2}\right)^2 \cdot \lambda \cdot \frac{-2\lambda r + a^2\sin^2\vartheta}{r^2 - 2\lambda r + a^2} + \left(\frac{\partial u^2}{\partial x^2}\right)^2 \cdot \lambda a^2\cos^2\vartheta$$

$$+ \left(\frac{\partial u^3}{\partial x^2}\right)^2 \cdot \left(-\lambda a^2\sin^2\vartheta - \lambda\sin^2\vartheta \cdot \frac{2\lambda r a^2\sin^2\vartheta}{r^2 + a^2\cos^2\vartheta}\right)$$

$$= \frac{\lambda(-2\lambda r + a^2\sin^2\vartheta)\sin^2\vartheta\sin^2\varphi}{r^2 - 2\lambda r + a^2} + \frac{\lambda a^2\cos^4\vartheta}{r^2\sin^2\varphi}$$

$$+ \frac{\sin^2\vartheta\cos^2\varphi}{r^2} \cdot \left(-\lambda a^2\sin^2\vartheta - \lambda\sin^2\vartheta \cdot \frac{2\lambda r a^2\sin^2\vartheta}{r^2 + a^2\cos^2\vartheta}\right) = \mathcal{O}\left(\frac{1}{r}\right)$$

$$g_{33}^K - g_{33}^M = \left(\frac{\partial u^1}{\partial x^3}\right)^2 \cdot \lambda \cdot \frac{-2\lambda r + a^2\sin^2\vartheta}{r^2 - 2\lambda r + a^2} + \left(\frac{\partial u^2}{\partial x^3}\right)^2 \cdot \lambda a^2\cos^2\vartheta$$

$$+ \left(\frac{\partial u^3}{\partial x^3}\right)^2 \cdot \left(-\lambda a^2\sin^2\vartheta - \lambda\sin^2\vartheta \cdot \frac{2\lambda r a^2\sin^2\vartheta}{r^2 + a^2\cos^2\vartheta}\right)$$

$$= \frac{\lambda(-2\lambda r + a^2\sin^2\vartheta)\cos^2\varphi}{r^2 - 2\lambda r + a^2} + 0$$

$$+ \frac{\sin^2\varphi}{r^2} \cdot \left(-\lambda a^2\sin^2\vartheta - \lambda\sin^2\vartheta \cdot \frac{2\lambda r a^2\sin^2\vartheta}{r^2 + a^2\cos^2\vartheta}\right) = \mathcal{O}\left(\frac{1}{r}\right).$$

Thus, the components of the metric have the right vanishing behaviour. Now we have to consider the Christoffel symbols. The following holds:

$$|{}^K\Gamma^1_{00} - {}^M\Gamma^1_{00}| = \left|\frac{(r^2 - 2\lambda r + a^2)(r^2 - a^2\cos^2\vartheta)}{(r^2 + a^2\cos^2\vartheta)^3}\right| = \mathcal{O}\left(\frac{1}{r^2}\right)$$

$$\left|{}^{K}\Gamma_{00}^{2} - {}^{M}\Gamma_{00}^{2}\right| = \left|\frac{-2ra^{2}\sin\vartheta\cos\vartheta}{(r^{2}+a^{2}\cos^{2}\vartheta)^{3}}\right| = \mathcal{O}\left(\frac{1}{r^{5}}\right)$$

$$\left|{}^{K}\Gamma_{01}^{0} - {}^{M}\Gamma_{01}^{0}\right| = \left|\frac{\lambda(r^{2}-a^{2}\cos^{2}\vartheta)(r^{2}+a^{2})}{(r^{2}+a^{2}\cos^{2}\vartheta)^{2}(r^{2}-2\lambda r+a^{2})}\right| = \mathcal{O}\left(\frac{1}{r^{2}}\right)$$

$$\left|{}^{K}\Gamma_{02}^{0} - {}^{M}\Gamma_{02}^{0}\right| = \left|\frac{-4\lambda ra^{2}\sin\vartheta\cos\vartheta}{(r^{2}+a^{2}\cos^{2}\vartheta)^{2}}\right| = \mathcal{O}\left(\frac{1}{r^{3}}\right)$$

$$\left|{}^{K}\Gamma_{01}^{3} - {}^{M}\Gamma_{01}^{3}\right| = \left|\frac{\sqrt{\lambda}a(r^{2}-a^{2}\cos^{2}\vartheta)}{(r^{2}-2\lambda r+a^{2})(r^{2}+a^{2}\cos^{2}\vartheta)^{2}}\right| = \mathcal{O}\left(\frac{1}{r^{4}}\right)$$

$$\left|{}^{K}\Gamma_{02}^{3} - {}^{M}\Gamma_{02}^{3}\right| = \left|\frac{-2\sqrt{\lambda}ra\cos\vartheta}{\sin\vartheta\cdot(r^{2}+a^{2}\cos^{2}\vartheta)^{2}}\right| = \mathcal{O}\left(\frac{1}{r^{3}}\right)$$

$$\left|{}^{K}\Gamma_{13}^{0} - {}^{M}\Gamma_{13}^{0}\right| = \left|\frac{-\lambda^{\frac{3}{2}}a\sin^{2}\vartheta[(r^{2}+a^{2})(r^{2}-a^{2}\cos^{2}\vartheta)+2r^{2}\rho]}{(r^{2}+a^{2}\cos^{2}\vartheta)^{2}(r^{2}-2\lambda r+a^{2})}\right| = \mathcal{O}\left(\frac{1}{r^{3}}\right)$$

$$\left|{}^{K}\Gamma_{23}^{0} - {}^{M}\Gamma_{23}^{0}\right| = \left|\frac{2\lambda^{\frac{3}{2}}ra^{3}\sin^{3}\vartheta\cos\vartheta}{(r^{2}+a^{2}\cos^{2}\vartheta)^{2}}\right| = \mathcal{O}\left(\frac{1}{r^{3}}\right)$$

$$\left|{}^{K}\Gamma_{30}^{1} - {}^{M}\Gamma_{30}^{1}\right| = \left|-\frac{\sqrt{\lambda}(r^{2}-2\lambda r+a^{2})\cdot a\sin^{2}\vartheta(r^{2}-a^{2}\cos^{2}\vartheta)}{(r^{2}+a^{2}\cos^{2}\vartheta)^{3}}\right| = \mathcal{O}\left(\frac{1}{r^{2}}\right)$$

$$\left|{}^{K}\Gamma_{30}^{2} - {}^{M}\Gamma_{30}^{2}\right| = \left|\frac{2\sqrt{\lambda}ra\sin\vartheta\cos\vartheta(r^{2}+a^{2})}{(r^{2}+a^{2}\cos^{2}\vartheta)^{3}}\right| = \mathcal{O}\left(\frac{1}{r^{3}}\right)$$

$$\left|{}^{K}\Gamma_{11}^{1} - {}^{M}\Gamma_{11}^{1}\right| = \left|\frac{ra^{2}\sin^{2}\vartheta-\lambda(r^{2}-a^{2}\cos^{2}\vartheta)}{(r^{2}+a^{2}\cos^{2}\vartheta)(r^{2}-2\lambda r+a^{2})}\right| = \mathcal{O}\left(\frac{1}{r^{2}}\right)$$

$$\left|{}^{K}\Gamma_{11}^{2} - {}^{M}\Gamma_{11}^{2}\right| = \left|\frac{a^{2}\sin\vartheta\cos\vartheta}{(r^{2}+a^{2}\cos^{2}\vartheta)(r^{2}-2\lambda r+a^{2})}\right| = \mathcal{O}\left(\frac{1}{r^{4}}\right)$$

$$\left|{}^{K}\Gamma_{21}^{1} - {}^{M}\Gamma_{21}^{1}\right| = \left|\frac{-a^{2}\sin\vartheta\cos\vartheta}{(r^{2}+a^{2}\cos^{2}\vartheta)}\right| = \mathcal{O}\left(\frac{1}{r^{2}}\right)$$

$$\left|{}^{K}\Gamma_{22}^{1} - {}^{M}\Gamma_{22}^{1}\right| = \left|-\frac{(r^{2}-2\lambda r+a^{2})r}{(r^{2}+a^{2}\cos^{2}\vartheta)} + \frac{r(r^{2}+a^{2}\cos^{2}\vartheta)}{r^{2}+a^{2}\cos^{2}\vartheta}\right|$$

$$= \left|\frac{2\lambda r^{2}-ra^{2}+ra^{2}\cos^{2}\vartheta}{r^{2}+a^{2}\cos^{2}\vartheta}\right| = 2\lambda + \mathcal{O}\left(\frac{1}{r}\right)$$

$$\left|{}^{K}\Gamma_{21}^{2} - {}^{M}\Gamma_{21}^{2}\right| = \left|\frac{r^{2}}{r(r^{2}+a^{2}\cos^{2}\vartheta)} - \frac{r^{2}+a^{2}\cos^{2}\vartheta}{r(r^{2}+a^{2}\cos^{2}\vartheta)}\right|$$

$$= \left|\frac{-a^{2}\cos^{2}\vartheta}{r^{3}+ra^{2}\cos^{2}\vartheta}\right| = \mathcal{O}\left(\frac{1}{r^{3}}\right)$$

$$\left|{}^{K}\Gamma_{22}^{2} - {}^{M}\Gamma_{22}^{2}\right| = \left|-\frac{a^{2}\cos\vartheta\sin\vartheta}{(r^{2}+a^{2}\cos^{2}\vartheta)}\right| = \mathcal{O}\left(\frac{1}{r^{2}}\right)$$

$$\left|{}^{K}\Gamma_{33}^{1} - {}^{M}\Gamma_{33}^{1}\right| = \left|-\frac{(r^{2}-2\lambda r+a^{2})\sin^{2}\vartheta(r(r^{2}+a^{2}\cos^{2}\vartheta)^{2}-\lambda a^{2}\sin^{2}\vartheta(r^{2}-a^{2}\cos\vartheta))}{(r^{2}+a^{2}\cos^{2}\vartheta)^{3}}\right.$$

$$\left. + \frac{r\sin^{2}\vartheta(r^{2}+a^{2}\cos^{2}\vartheta)^{3}}{(r^{2}+a^{2}\cos^{2}\vartheta)^{3}}\right|$$

$$= \left|\frac{-r^{7}\sin^{2}\vartheta+r^{7}\sin^{2}\vartheta+2\lambda\sin^{2}\vartheta r^{6}+\mathcal{O}(r^{5})}{r^{6}+\mathcal{O}(r^{4})}\right| = 2\lambda\sin^{2}\vartheta + \mathcal{O}\left(\frac{1}{r}\right)$$

$$|{}^K\Gamma^2_{33} - {}^M\Gamma^2_{33}| = \left|-\frac{\sin\vartheta\cos\vartheta((r^2-2\lambda r+a^2)(r^2+a^2\cos^2\vartheta)^2+2\lambda r(r^2+a^2)^2)}{(r^2+a^2\cos^2\vartheta)^3}\right|$$

$$= \left|\frac{-r^6\sin\vartheta\cos\vartheta - 2\lambda r^5 + 2\lambda r^5 + r^6\sin\vartheta\cos\vartheta + \mathcal{O}(r^4)}{\mathcal{O}(r^6)}\right| = \mathcal{O}\left(\frac{1}{r^2}\right)$$

$$|{}^K\Gamma^3_{31} - {}^M\Gamma^3_{31}| = \left|\frac{-\lambda r(2r^2(r^2+a^2\cos^2\vartheta)^2 - (r^2+a^2\cos^2\vartheta)a^2\sin^2\vartheta(r^2-a^2\cos\vartheta))}{r(r^2-2\lambda r+a^2)(r^2+a^2\cos^2\vartheta)^2}\right.$$

$$\left.+\frac{r^2(r^2+a^2\cos^2\vartheta)^2}{(r^2-2\lambda r+a^2)(r^2+a^2\cos^2\vartheta)^2} - \frac{(r^2+a^2\cos^2\vartheta)^2(r^2-2\lambda r+a^2)}{r(r^2+a^2\cos^2\vartheta)^2(r^2-2\lambda r+a^2)}\right|$$

$$= \left|\frac{r^6-r^6+\mathcal{O}(r^5)}{\mathcal{O}(r^7)}\right| = \mathcal{O}\left(\frac{1}{r^2}\right)$$

$$|{}^K\Gamma^3_{32} - {}^M\Gamma^3_{32}| = \left|\frac{\cos\vartheta(2\lambda ra^2\sin^2\vartheta + (r^2+a^2\cos^2\vartheta)^2)}{\sin\vartheta\cdot(r^2+a^2\cos^2\vartheta)^2} - \frac{\cos\vartheta(r^2+a^2\cos^2\vartheta)^2}{\sin\vartheta(r^2+a^2\cos^2\vartheta)^2}\right|$$

$$= \left|\frac{2\lambda ra^2\sin\vartheta\cos\vartheta}{(r^2+a^2\cos^2\vartheta)^2}\right| = \mathcal{O}\left(\frac{1}{r^3}\right).$$

Again we see that at two points the condition is not satisfied. Therefore, we have to consider the transformation behaviour. As we have already seen only $\left(\frac{\partial x}{\partial u}\right)$ has components of order r. Since we just have to multiply the Christoffel symbols with one component of this matrix, we do not have to consider the differences which are of order $\left(\frac{1}{r^3}\right)$ or higher powers. As the two first columns of $\left(\frac{\partial x}{\partial u}\right)$ do not depend on r, we do not have to care about the differences of order $\left(\frac{1}{r^2}\right)$ with $p=0$ or $p=1$. So we just have to discuss the following cases:

(i)
$$\Gamma^1_{22}: \frac{\partial u^2}{\partial x^\alpha}\cdot\frac{\partial u^2}{\partial x^\beta}\cdot\frac{\partial x^\gamma}{\partial u^1}({}^K\Gamma^1_{22} - {}^M\Gamma^1_{22})$$

for $\alpha,\beta,\gamma\in\{0,1,2,3\}$. As $p=1$, the behaviour does not get worse. For $\alpha,\beta\in\{0,3\}$ the Christoffel symbol equals zero, for $\alpha,\beta\in\{1,2\}$ it is of order $\left(\frac{1}{r^2}\right)$.

(ii)
$$\Gamma^2_{22}: \frac{\partial u^2}{\partial x^\alpha}\cdot\frac{\partial u^2}{\partial x^\beta}\cdot\frac{\partial x^\gamma}{\partial u^2}({}^K\Gamma^2_{22} - {}^M\Gamma^2_{22})$$

for $\alpha,\beta,\gamma\in\{0,1,2,3\}$. In this case $p=2$, so we just have to show that it does not deteriorate: ${}^K\Gamma^2_{22} - {}^M\Gamma^2_{22}$ is of order $\left(\frac{1}{r^2}\right)$ and for $\alpha,\beta\in\{0,3\}$ it equals zero. For $\alpha,\beta\in\{1,2\}$ we get a $\mathcal{O}(r)-$ and an $\mathcal{O}\left(\frac{1}{r^2}\right)-$ term. Altogether, it then is of order $\left(\frac{1}{r^3}\right)$.

For the other critical cases we always have to use the same arguments. For Γ^1_{33} we again have case (i), for Γ^2_{33} and Γ^3_{31} we use case (ii). But this already shows that the vanishing behaviour of the Kerr Christoffel symbols is of order $\left(\frac{1}{r^2}\right)$ which shows that the extended Kerr spacetime meets the conditions of definition (3.3.8) since now all transformed Christoffel symbols are of order $\left(\frac{1}{r^2}\right)$ as sums of $\mathcal{O}\left(\frac{1}{r^2}\right)$−terms.

Now we have a look at the components of the curvature tensor. Again we are interested in the vanishing behaviour after transformation. As before, for R^l_{ijk} we can get a factor of order r, if $l=2$ or $l=3$ (or the component already equals 0) and a factor of order

$\left(\frac{1}{r}\right)$, if i, j or k equal 2 or 3. The orders before transformation are calculated in (A.2.5). The orders after transformation can be found in the following table:

component	order before transformation	number of $\mathcal{O}(r)$-factors	number of $\mathcal{O}\left(\frac{1}{r}\right)$-factors	order after transformation
R^0_{030}	$\mathcal{O}\left(\frac{1}{r^4}\right)$	0	1	$\mathcal{O}\left(\frac{1}{r^5}\right)$
R^0_{110}	$\mathcal{O}\left(\frac{1}{r^3}\right)$	0	0	$\mathcal{O}\left(\frac{1}{r^3}\right)$
R^0_{220}	$\mathcal{O}\left(\frac{1}{r}\right)$	0	2	$\mathcal{O}\left(\frac{1}{r^3}\right)$
R^0_{330}	$\mathcal{O}\left(\frac{1}{r}\right)$	0	2	$\mathcal{O}\left(\frac{1}{r^3}\right)$
R^0_{120}	$\mathcal{O}\left(\frac{1}{r^4}\right)$	0	1	$\mathcal{O}\left(\frac{1}{r^5}\right)$
R^0_{210}	$\mathcal{O}\left(\frac{1}{r^4}\right)$	0	1	$\mathcal{O}\left(\frac{1}{r^5}\right)$
R^0_{012}	$\mathcal{O}\left(\frac{1}{r^4}\right)$	0	1	$\mathcal{O}\left(\frac{1}{r^5}\right)$
R^1_{010}	$\mathcal{O}\left(\frac{1}{r^3}\right)$	0	0	$\mathcal{O}\left(\frac{1}{r^3}\right)$
R^2_{020}	$\mathcal{O}\left(\frac{1}{r^3}\right)$	1	1	$\mathcal{O}\left(\frac{1}{r^3}\right)$
R^3_{030}	$\mathcal{O}\left(\frac{1}{r^3}\right)$	1	1	$\mathcal{O}\left(\frac{1}{r^3}\right)$
R^1_{020}	$\mathcal{O}\left(\frac{1}{r^4}\right)$	0	1	$\mathcal{O}\left(\frac{1}{r^5}\right)$
R^2_{010}	$\mathcal{O}\left(\frac{1}{r^6}\right)$	1	0	$\mathcal{O}\left(\frac{1}{r^5}\right)$
R^0_{113}	$\mathcal{O}\left(\frac{1}{r^3}\right)$	0	1	$\mathcal{O}\left(\frac{1}{r^4}\right)$
R^3_{110}	$\mathcal{O}\left(\frac{1}{r^5}\right)$	1	0	$\mathcal{O}\left(\frac{1}{r^4}\right)$
R^1_{310}	$\mathcal{O}\left(\frac{1}{r^3}\right)$	0	1	$\mathcal{O}\left(\frac{1}{r^4}\right)$
R^1_{130}	$\mathcal{O}\left(\frac{1}{r^3}\right)$	0	1	$\mathcal{O}\left(\frac{1}{r^4}\right)$
R^1_{013}	$\mathcal{O}\left(\frac{1}{r^3}\right)$	0	1	$\mathcal{O}\left(\frac{1}{r^4}\right)$
R^0_{223}	$\mathcal{O}\left(\frac{1}{r}\right)$	0	3	$\mathcal{O}\left(\frac{1}{r^4}\right)$
R^3_{220}	$\mathcal{O}\left(\frac{1}{r^3}\right)$	1	2	$\mathcal{O}\left(\frac{1}{r^4}\right)$
R^2_{320}	$\mathcal{O}\left(\frac{1}{r^3}\right)$	1	2	$\mathcal{O}\left(\frac{1}{r^4}\right)$
R^2_{230}	$\mathcal{O}\left(\frac{1}{r^3}\right)$	1	2	$\mathcal{O}\left(\frac{1}{r^4}\right)$
R^2_{023}	$\mathcal{O}\left(\frac{1}{r^3}\right)$	1	2	$\mathcal{O}\left(\frac{1}{r^4}\right)$
R^0_{123}	$\mathcal{O}\left(\frac{1}{r^2}\right)$	0	2	$\mathcal{O}\left(\frac{1}{r^4}\right)$

3.3. Asymptotically flat spacetimes

component	order before transformation	number of $\mathcal{O}(r)$-factors	number of $\mathcal{O}\left(\frac{1}{r}\right)$-factors	order after transformation
R^0_{213}	$\mathcal{O}\left(\frac{1}{r}\right)$	0	2	$\mathcal{O}\left(\frac{1}{r^3}\right)$
R^0_{312}	$\mathcal{O}\left(\frac{1}{r^2}\right)$	0	2	$\mathcal{O}\left(\frac{1}{r^4}\right)$
R^3_{012}	$\mathcal{O}\left(\frac{1}{r^4}\right)$	1	1	$\mathcal{O}\left(\frac{1}{r^4}\right)$
R^3_{102}	$\mathcal{O}\left(\frac{1}{r^4}\right)$	1	1	$\mathcal{O}\left(\frac{1}{r^4}\right)$
R^3_{201}	$\mathcal{O}\left(\frac{1}{r^4}\right)$	1	1	$\mathcal{O}\left(\frac{1}{r^4}\right)$
R^1_{023}	$\mathcal{O}\left(\frac{1}{r^2}\right)$	0	2	$\mathcal{O}\left(\frac{1}{r^4}\right)$
R^1_{203}	$\mathcal{O}\left(\frac{1}{r^2}\right)$	0	2	$\mathcal{O}\left(\frac{1}{r^4}\right)$
R^1_{302}	$\mathcal{O}\left(\frac{1}{r^2}\right)$	0	2	$\mathcal{O}\left(\frac{1}{r^4}\right)$
R^2_{013}	$\mathcal{O}\left(\frac{1}{r^4}\right)$	1	1	$\mathcal{O}\left(\frac{1}{r^4}\right)$
R^2_{103}	$\mathcal{O}\left(\frac{1}{r^4}\right)$	1	1	$\mathcal{O}\left(\frac{1}{r^4}\right)$
R^2_{310}	$\mathcal{O}\left(\frac{1}{r^4}\right)$	1	1	$\mathcal{O}\left(\frac{1}{r^4}\right)$
R^3_{330}	$\mathcal{O}\left(\frac{1}{r^3}\right)$	1	2	$\mathcal{O}\left(\frac{1}{r^4}\right)$
R^1_{112}	$\mathcal{O}\left(\frac{1}{r^3}\right)$	0	1	$\mathcal{O}\left(\frac{1}{r^4}\right)$
R^2_{221}	$\mathcal{O}\left(\frac{1}{r^3}\right)$	1	2	$\mathcal{O}\left(\frac{1}{r^4}\right)$
R^1_{212}	$\mathcal{O}\left(\frac{1}{r}\right)$	0	2	$\mathcal{O}\left(\frac{1}{r^3}\right)$
R^2_{121}	$\mathcal{O}\left(\frac{1}{r^3}\right)$	1	1	$\mathcal{O}\left(\frac{1}{r^3}\right)$
R^1_{313}	$\mathcal{O}\left(\frac{1}{r}\right)$	0	2	$\mathcal{O}\left(\frac{1}{r^3}\right)$
R^3_{131}	$\mathcal{O}\left(\frac{1}{r^3}\right)$	1	1	$\mathcal{O}\left(\frac{1}{r^3}\right)$
R^2_{323}	$\mathcal{O}\left(\frac{1}{r}\right)$	1	3	$\mathcal{O}\left(\frac{1}{r^3}\right)$
R^3_{232}	$\mathcal{O}\left(\frac{1}{r}\right)$	1	3	$\mathcal{O}\left(\frac{1}{r^3}\right)$
R^3_{312}	$\mathcal{O}\left(\frac{1}{r^4}\right)$	1	2	$\mathcal{O}\left(\frac{1}{r^5}\right)$
R^3_{132}	$\mathcal{O}\left(\frac{1}{r^2}\right)$	1	2	$\mathcal{O}\left(\frac{1}{r^3}\right)$
R^3_{231}	$\mathcal{O}\left(\frac{1}{r^2}\right)$	1	2	$\mathcal{O}\left(\frac{1}{r^3}\right)$
R^1_{332}	const.	0	3	$\mathcal{O}\left(\frac{1}{r^3}\right)$
R^2_{331}	$\mathcal{O}\left(\frac{1}{r^2}\right)$	1	2	$\mathcal{O}\left(\frac{1}{r^3}\right)$

Now we can see that the components of the curvature tensor have the right vanishing behaviour after transformation. As in the case of Schwarzschild, we can also exchange the limits for $\lambda \to 0$ and $r \to \infty$ in the Christoffel symbols. Therefore, the extended Kerr spacetime is an example of an asymptotically flat family of Ehlers spacetimes.

Thus, we can say that our standard extended examples satisfy the definition of asymptotically flat families. This is a very important confirmation of our definition.

Appendix A

Appendix

A.1 The Schwarzschild spacetime

In Appendix A.1 we are going to discuss the calculations for the Schwarzschild solution more closely.

The Christoffel symbols A.1.1 We first calculate the Christoffel symbols of the extended Schwarzschild solution:

Generally, the following equations hold for a metric in diagonal form:

$$\Gamma^i_{ii} = \frac{1}{2} g^{ii} \partial_i g_{ii}, \quad i = 0, ..., 3;$$

$$\Gamma^k_{ii} = -\frac{1}{2} g^{kk} \partial_k g_{ii}, \quad k \neq i;$$

$$\Gamma^j_{ij} = \frac{1}{2} g^{jj} \partial_i g_{jj}, \quad j \neq i;$$

$$\Gamma^k_{ij} = 0, \quad \text{otherwise.}$$

Now we have to consider the following

$$(g_{ij}(\lambda)) = \text{diag}\left(1 - \frac{2\lambda}{r}, \frac{-r\lambda}{r - 2\lambda}, -\lambda r^2, -\lambda r^2 \sin^2 \vartheta\right),$$

$$(g^{ij}(\lambda)) = \text{diag}\left(\frac{r}{r - 2\lambda}, \frac{r - 2\lambda}{-r\lambda}, -\frac{1}{\lambda r^2}, -\frac{1}{\lambda r^2 \sin^2 \vartheta}\right).$$

The Christoffel symbols have the form (with $\partial_0 = \partial_t$, $\partial_1 = \partial_r$, $\partial_2 = \partial_\vartheta$ and $\partial_3 = \partial_\varphi$):

$$\Gamma^0_{00} = 0$$

$$\Gamma^1_{11} = \frac{1}{2}\left(\frac{r - 2\lambda}{-r\lambda}\right) \partial_r \left(\frac{-r\lambda}{r - 2\lambda}\right) = \frac{1}{2}\left(\frac{r - 2\lambda}{-r\lambda}\right)\left(\frac{-\lambda r + 2\lambda^2 + \lambda r}{(r - 2\lambda)^2}\right) = \frac{\lambda}{2\lambda r - r^2}$$

$$\Gamma^2_{22} = 0$$

$$\Gamma^3_{33} = 0$$

$$\Gamma^1_{00} = -\frac{1}{2}\left(\frac{r - 2\lambda}{-r\lambda}\right) \partial_r \left(1 - \frac{2\lambda}{r}\right) = -\frac{1}{2}\left(\frac{r - 2\lambda}{-r\lambda}\right)\left(\frac{2\lambda}{r^2}\right) = \frac{r - 2\lambda}{r^3}$$

A.1. The Schwarzschild spacetime

$$\Gamma^2_{00} = 0$$
$$\Gamma^3_{00} = 0$$
$$\Gamma^0_{11} = 0$$
$$\Gamma^2_{11} = 0$$
$$\Gamma^3_{11} = 0$$
$$\Gamma^0_{22} = 0$$
$$\Gamma^1_{22} = -\frac{1}{2}\left(\frac{r-2\lambda}{-r\lambda}\right)\partial_r(-\lambda r^2) = -\frac{1}{2}\left(\frac{r-2\lambda}{-r\lambda}\right)(-2\lambda r) = 2\lambda - r$$
$$\Gamma^3_{22} = 0$$
$$\Gamma^0_{33} = 0$$
$$\Gamma^1_{33} = -\frac{1}{2}\left(\frac{r-2\lambda}{-r\lambda}\right)\partial_r(-\lambda r^2 \sin^2\vartheta) = \frac{1}{2}\left(\frac{r-2\lambda}{r\lambda}\right)(-2\lambda r \sin^2\vartheta) = (2\lambda - r)\sin^2\vartheta$$
$$\Gamma^2_{33} = -\frac{1}{2}\left(-\frac{1}{\lambda r^2}\right)\partial_\vartheta(-\lambda r^2 \sin^2\vartheta) = \frac{1}{2}\left(\frac{1}{\lambda r^2}\right)(-2\lambda r^2 \sin\vartheta\cos\vartheta) = -\sin\vartheta\cos\vartheta$$
$$\Gamma^1_{01} = 0$$
$$\Gamma^2_{02} = 0$$
$$\Gamma^3_{03} = 0$$
$$\Gamma^0_{10} = \frac{1}{2}\left(\frac{r}{r-2\lambda}\right)\partial_r\left(1-\frac{2\lambda}{r}\right) = \frac{1}{2}\left(\frac{r}{r-2\lambda}\right)\left(\frac{2\lambda}{r^2}\right) = \frac{\lambda}{r^2 - 2\lambda r}$$
$$\Gamma^2_{12} = \frac{1}{2}\left(-\frac{1}{\lambda r^2}\right)\partial_r(-\lambda r^2) = \frac{1}{2}\left(-\frac{1}{\lambda r^2}\right)(-2\lambda r) = \frac{1}{r}$$
$$\Gamma^3_{13} = \frac{1}{2}\left(-\frac{1}{\lambda r^2 \sin^2\vartheta}\right)\partial_r(-\lambda r^2 \sin^2\vartheta) = \frac{1}{2}\left(-\frac{1}{\lambda r^2 \sin^2\vartheta}\right)(-2\lambda r \sin^2\vartheta) = \frac{1}{r}$$
$$\Gamma^0_{20} = 0$$
$$\Gamma^1_{21} = 0$$
$$\Gamma^3_{23} = \frac{1}{2}\left(-\frac{1}{\lambda r^2 \sin^2\vartheta}\right)\partial_\vartheta(-\lambda r^2 \sin^2\vartheta) = \frac{1}{2}\left(-\frac{1}{\lambda r^2 \sin^2\vartheta}\right)(-2\lambda r^2 \sin\vartheta\cos\vartheta) = \frac{\cos\vartheta}{\sin\vartheta}$$
$$\Gamma^0_{30} = 0$$
$$\Gamma^1_{31} = 0$$
$$\Gamma^2_{32} = 0.$$

The Christoffel symbols of the limit connection A.1.2 The Christoffel symbols of the limit connection take the form

$$\Gamma^1_{00} = \frac{1}{r^2}$$
$$\Gamma^1_{22} = -r$$
$$\Gamma^1_{33} = -r\sin^2\vartheta$$
$$\Gamma^2_{33} = -\sin\vartheta\cos\vartheta$$

$$\Gamma_{12}^2 = \frac{1}{r}$$
$$\Gamma_{13}^3 = \frac{1}{r}$$
$$\Gamma_{23}^3 = \frac{\cos\vartheta}{\sin\vartheta}$$
$$\Gamma_{ij}^k = 0, \text{ otherwise.}$$

The components of the Ricci tensor of the limit A.1.3 In the following we calculate the components of the Ricci tensor. As it is symmetric we only have to study the following components:

$$\text{Ric}_{00} = \partial_j \Gamma_{00}^j - \partial_0 \Gamma_{j0}^j + \Gamma_{jr}^j \Gamma_{00}^r - \Gamma_{0r}^j \Gamma_{j0}^r = \partial_r\left(\frac{1}{r^2}\right) - 0 + \frac{1}{r} \cdot \frac{1}{r^2} + \frac{1}{r} \cdot \frac{1}{r^2} - 0 = -\frac{2}{r^3} + \frac{2}{r^3} = 0$$

$$\text{Ric}_{01} = \partial_j \Gamma_{10}^j - \partial_1 \Gamma_{j0}^j + \Gamma_{jr}^j \Gamma_{10}^r - \Gamma_{1r}^j \Gamma_{j0}^r = 0$$

$$\text{Ric}_{02} = \partial_j \Gamma_{20}^j - \partial_2 \Gamma_{j0}^j + \Gamma_{jr}^j \Gamma_{20}^r - \Gamma_{2r}^j \Gamma_{j0}^r = 0$$

$$\text{Ric}_{03} = \partial_j \Gamma_{30}^j - \partial_3 \Gamma_{j0}^j + \Gamma_{jr}^j \Gamma_{30}^r - \Gamma_{3r}^j \Gamma_{j0}^r = 0$$

$$\text{Ric}_{11} = \partial_j \Gamma_{11}^j - \partial_1 \Gamma_{j1}^j + \Gamma_{jr}^j \Gamma_{11}^r - \Gamma_{1r}^j \Gamma_{j1}^r = 0 - \partial_r\left(\frac{1}{r}\right) - \partial_r\left(\frac{1}{r}\right) + 0 - \frac{1}{r^2} - \frac{1}{r^2}$$
$$= \frac{2}{r^2} - \frac{2}{r^2} = 0$$

$$\text{Ric}_{12} = \partial_j \Gamma_{21}^j - \partial_2 \Gamma_{j1}^j + \Gamma_{jr}^j \Gamma_{21}^r - \Gamma_{2r}^j \Gamma_{j1}^r = 0 - 0 + \frac{\cos\vartheta}{\sin\vartheta} \cdot \frac{1}{r} - \frac{1}{r} \cdot \frac{\cos\vartheta}{\sin\vartheta} = 0$$

$$\text{Ric}_{13} = \partial_j \Gamma_{31}^j - \partial_3 \Gamma_{j1}^j + \Gamma_{jr}^j \Gamma_{31}^r - \Gamma_{3r}^j \Gamma_{j1}^r = 0$$

$$\text{Ric}_{22} = \partial_j \Gamma_{22}^j - \partial_2 \Gamma_{j2}^j + \Gamma_{jr}^j \Gamma_{22}^r - \Gamma_{2r}^j \Gamma_{j2}^r$$
$$= \partial_1 \Gamma_{22}^1 - \partial_2 \Gamma_{32}^3 + \Gamma_{21}^2 \Gamma_{22}^1 + \Gamma_{31}^3 \Gamma_{22}^1 - \Gamma_{21}^2 \Gamma_{22}^1 - \Gamma_{22}^1 \Gamma_{12}^2 - \Gamma_{23}^3 \Gamma_{32}^3$$
$$= \partial_1 \Gamma_{22}^1 - \partial_2 \Gamma_{32}^3 + \Gamma_{31}^3 \Gamma_{22}^1 - \Gamma_{22}^1 \Gamma_{12}^2 - \Gamma_{23}^3 \Gamma_{32}^3$$
$$= \partial_r(-r) - \partial_\vartheta\left(\frac{\cos\vartheta}{\sin\vartheta}\right) + \frac{1}{r} \cdot (-r) - (-r) \cdot \frac{1}{r} - \frac{\cos^2\vartheta}{\sin^2\vartheta}$$
$$= -1 + \frac{\sin^2\vartheta + \cos^2\vartheta}{\sin^2\vartheta} - 1 + 1 - \frac{\cos^2\vartheta}{\sin^2\vartheta} = -1 + \frac{1-\cos^2\vartheta}{\sin^2\vartheta} = 0$$

$$\text{Ric}_{23} = \partial_j \Gamma_{32}^j - \partial_3 \Gamma_{j2}^j + \Gamma_{jr}^j \Gamma_{32}^r - \Gamma_{3r}^j \Gamma_{j2}^r = 0$$

$$\text{Ric}_{33} = \partial_j \Gamma_{33}^j - \partial_3 \Gamma_{j3}^j + \Gamma_{jr}^j \Gamma_{33}^r - \Gamma_{3r}^j \Gamma_{j3}^r$$
$$= \partial_1 \Gamma_{33}^1 + \partial_2 \Gamma_{33}^2 + \Gamma_{21}^2 \Gamma_{33}^1 + \Gamma_{31}^3 \Gamma_{33}^1 + \Gamma_{32}^3 \Gamma_{33}^2 - \Gamma_{33}^1 \Gamma_{13}^3 - \Gamma_{33}^2 \Gamma_{23}^3 - \Gamma_{31}^3 \Gamma_{33}^1 - \Gamma_{32}^3 \Gamma_{33}^2$$
$$= \partial_1 \Gamma_{33}^1 + \partial_2 \Gamma_{33}^2 + \Gamma_{21}^2 \Gamma_{33}^1 - \Gamma_{33}^1 \Gamma_{13}^3 - \Gamma_{33}^2 \Gamma_{23}^3$$
$$= \partial_r(-r\sin^2\vartheta) + \partial_\vartheta(-\sin\vartheta\cos\vartheta) + \frac{1}{r} \cdot (-r\sin\vartheta) - (-r\sin^2\vartheta) \cdot \frac{1}{r}$$
$$- (-\sin\vartheta\cos\vartheta) \cdot \frac{\cos\vartheta}{\sin\vartheta}$$
$$= -\sin^2\vartheta - \cos^2\vartheta + \sin^2\vartheta - \sin^2\vartheta + \sin^2\vartheta + \cos^2\vartheta = 0.$$

The divergence of the function f A.1.4 Since the Ricci tensor of the limit of Schwarzschild spacetime vanishes, we already know that $\rho = 0$. Therefore, we now have to show that the

A.1. The Schwarzschild spacetime

equation $\Delta u = 0$ holds. As we already have $f = \text{grad}(u)$, we just have to demonstrate that $\text{div}(f) = 0$. In cartesian coordinates $\text{div}(f)$ is given by

$$\text{div}(f) = \frac{\partial}{\partial x^1} f^1 + \frac{\partial}{\partial x^2} f^2 + \frac{\partial}{\partial x^3} f^3.$$

But in our case we use spherical coordinates. We therefore have to transform the components of the function and the differential operators. Since f is defined by the Christoffel symbols of the connection, its transformation behaviour is given by

$$\tilde{\Gamma}^k_{ij}(x(u)) = \left(\frac{\partial x^k}{\partial u^p} \cdot \frac{\partial u^m}{\partial x^i} \circ x \cdot \left(\frac{\partial^2 u^p}{\partial x^j \partial x^l} \circ x \cdot \frac{\partial x^l}{\partial u^m} + \frac{\partial u^n}{\partial x^j} \circ x \cdot \Gamma^p_{mn} \right) \right)(u).$$

Therefore, let $u = (t, r, \vartheta, \varphi)$ be spherical and $x = (t, x^1, x^2, x^3)$ cartesian coordinates. Then the transformation matrices have the form

$$\left(\frac{\partial x}{\partial u} \right) = \begin{pmatrix} 1 & 0 & 0 & 0 \\ 0 & \cos \vartheta \sin \varphi & -r \sin \vartheta \sin \varphi & r \cos \vartheta \cos \varphi \\ 0 & \sin \vartheta \sin \varphi & r \cos \vartheta \sin \varphi & r \sin \vartheta \cos \varphi \\ 0 & \cos \varphi & 0 & -r \sin \varphi \end{pmatrix},$$

and

$$\left(\frac{\partial u}{\partial x} \right) = \begin{pmatrix} 1 & 0 & 0 & 0 \\ 0 & \cos \vartheta \sin \varphi & \sin \vartheta \sin \varphi & \cos \varphi \\ 0 & -\frac{\sin \vartheta}{r \sin \varphi} & \frac{\cos \vartheta}{r \sin \varphi} & 0 \\ 0 & \frac{\cos \vartheta \cos \varphi}{r} & \frac{\sin \vartheta \cos \varphi}{r} & -\frac{\sin \varphi}{r} \end{pmatrix}.$$

In our case f is of the form

$$f = (\Gamma^1_{00}, \Gamma^2_{00}, \Gamma^3_{00}) = \left(\frac{1}{r^2}, 0, 0 \right).$$

We now have:

$$\tilde{\Gamma}^k_{00} = \frac{\partial x^k}{\partial u^p} \cdot \underbrace{\frac{\partial u^m}{\partial x^0}}_{\Rightarrow m=0} \left(\underbrace{\frac{\partial^2 u^p}{\partial x^0 \partial x^l} \cdot \frac{\partial x^l}{\partial u^m}}_{=0} + \underbrace{\frac{\partial u^n}{\partial x^0}}_{\Rightarrow n=0} \cdot \Gamma^p_{mn} \right) = \frac{\partial x^k}{\partial u^p} \cdot \Gamma^p_{00}.$$

Therefore, we get:

$$\tilde{\Gamma}^1_{00} = \cos \vartheta \sin \varphi \cdot \Gamma^1_{00} = \frac{\cos \vartheta \sin \varphi}{r^2}$$

$$\tilde{\Gamma}^2_{00} = \sin \vartheta \sin \varphi \cdot \Gamma^1_{00} = \frac{\sin \vartheta \sin \varphi}{r^2}$$

$$\tilde{\Gamma}^3_{00} = \cos \varphi \cdot \Gamma^1_{00} = \frac{\cos \varphi}{r^2}.$$

A.1. The Schwarzschild spacetime

For the differential operators we get:

$$\frac{\partial}{\partial x^1} = \frac{\partial r}{\partial x^1}\frac{\partial}{\partial r} + \frac{\partial \vartheta}{\partial x^1}\frac{\partial}{\partial \vartheta} + \frac{\partial \varphi}{\partial x^1}\frac{\partial}{\partial \varphi} = \cos\vartheta\sin\varphi\frac{\partial}{\partial r} - \frac{\sin\vartheta}{r\sin\varphi}\frac{\partial}{\partial \vartheta} + \frac{\cos\vartheta\cos\varphi}{r}\frac{\partial}{\partial \varphi}$$

$$\frac{\partial}{\partial x^2} = \frac{\partial r}{\partial x^2}\frac{\partial}{\partial r} + \frac{\partial \vartheta}{\partial x^2}\frac{\partial}{\partial \vartheta} + \frac{\partial \varphi}{\partial x^2}\frac{\partial}{\partial \varphi} = \sin\vartheta\sin\varphi\frac{\partial}{\partial r} + \frac{\cos\vartheta}{r\sin\varphi}\frac{\partial}{\partial \vartheta} + \frac{\sin\vartheta\cos\varphi}{r}\frac{\partial}{\partial \varphi}$$

$$\frac{\partial}{\partial x^3} = \frac{\partial r}{\partial x^3}\frac{\partial}{\partial r} + \frac{\partial \vartheta}{\partial x^3}\frac{\partial}{\partial \vartheta} + \frac{\partial \varphi}{\partial x^3}\frac{\partial}{\partial \varphi} = \cos\varphi\frac{\partial}{\partial r} - \sin\varphi\frac{\partial}{\partial \varphi}.$$

That is why the following holds for the divergence of f:

$$\begin{aligned}
\operatorname{div}(f) &= \frac{\partial}{\partial x^1}f^1 + \frac{\partial}{\partial x^2}f^2 + \frac{\partial}{\partial x^3}f^3 \\
&= \left(\cos\vartheta\sin\varphi\frac{\partial}{\partial r} - \frac{\sin\vartheta}{r\sin\varphi}\frac{\partial}{\partial \vartheta} + \frac{\cos\vartheta\cos\varphi}{r}\frac{\partial}{\partial \varphi}\right)\left(\frac{\cos\vartheta\sin\varphi}{r^2}\right) \\
&\quad + \left(\sin\vartheta\sin\varphi\frac{\partial}{\partial r} + \frac{\cos\vartheta}{r\sin\varphi}\frac{\partial}{\partial \vartheta} + \frac{\sin\vartheta\cos\varphi}{r}\frac{\partial}{\partial \varphi}\right)\left(\frac{\sin\vartheta\sin\varphi}{r^2}\right) \\
&\quad + \left(\cos\varphi\frac{\partial}{\partial r} - \sin\varphi\frac{\partial}{\partial \varphi}\right)\left(\frac{\cos\varphi}{r^2}\right) \\
&= \frac{-2\cos^2\vartheta\sin^2\varphi}{r^3} + \frac{\sin^2\vartheta}{r^3} + \frac{\cos^2\vartheta\cos^2\varphi}{r^3} - \frac{2\sin^2\vartheta\sin^2\varphi}{r^3} + \frac{\cos^2\vartheta}{r^3} \\
&\quad + \frac{\sin^2\vartheta\cos^2\varphi}{r^3} - \frac{2\cos^2\varphi}{r^3} + \frac{\sin^2\varphi}{r^3} \\
&= \frac{1}{r^3}(-2\sin^2\varphi(\cos^2\vartheta + \sin^2\vartheta) + \sin^2\vartheta + \cos^2\vartheta + \cos^2\varphi(\cos^2\vartheta + \sin^2\vartheta) \\
&\quad - 2\cos^2\varphi + \sin^2\varphi) = \frac{1}{r^3}(-\sin^2\varphi - \cos^2\varphi + 1) = 0.
\end{aligned}$$

But this is just what we wanted to show. Therefore, $\Delta u = 0$ holds.

The components of the curvature tensor A.1.5 We now calculate the components of the curvature tensor. The following holds:

$$R^l_{ijk} = \partial_j \Gamma^l_{ki} - \partial_k \Gamma^l_{ji} + \Gamma^l_{jr}\Gamma^r_{ki} - \Gamma^l_{kr}\Gamma^r_{ij}.$$

As the connection is symmetric,

$$R^l_{ijk} = -R^l_{ikj}$$

also holds. We only care for the components which do not vanish immediately. First, we have to calculate the derivations of the Christoffel symbols - of course, only for $i = 1$ (which means for r) and for $i = 2$ (which means for ϑ). We get:

$$\partial_1 \Gamma^1_{11} = \partial_1\left(\frac{\lambda}{2\lambda r - r^2}\right) = \frac{2\lambda(r-\lambda)}{(2\lambda r - r^2)^2}$$

$$\partial_1 \Gamma^1_{00} = \partial_1\left(\frac{r - 2\lambda}{r^3}\right) = \frac{6\lambda - 2r}{r^4}$$

$$\partial_1 \Gamma^1_{22} = \partial_1(2\lambda - r) = -1$$

A.1. The Schwarzschild spacetime

$$\partial_1 \Gamma^1_{33} = \partial_1((2\lambda - r)\sin^2\vartheta) = -\sin^2\vartheta$$

$$\partial_1 \Gamma^2_{33} = \partial_1(-\sin\vartheta\cos\vartheta) = 0$$

$$\partial_1 \Gamma^0_{10} = \partial_1\left(\frac{\lambda}{r^2 - 2\lambda r}\right) = \frac{2\lambda^2 - 2\lambda r}{(r^2 - 2\lambda r)^2}$$

$$\partial_1 \Gamma^2_{12} = \partial_1\left(\frac{1}{r}\right) = -\frac{1}{r^2}$$

$$\partial_1 \Gamma^3_{13} = \partial_1\left(\frac{1}{r}\right) = -\frac{1}{r^2}$$

$$\partial_1 \Gamma^3_{23} = \partial_1\left(\frac{\cos\vartheta}{\sin\vartheta}\right) = 0$$

$$\partial_2 \Gamma^1_{11} = \partial_2\left(\frac{r - 2\lambda}{r^3}\right) = 0$$

$$\partial_2 \Gamma^1_{00} = \partial_2\left(\frac{r - 2\lambda}{r^3}\right) = 0$$

$$\partial_2 \Gamma^1_{22} = \partial_2(2\lambda - r) = 0$$

$$\partial_2 \Gamma^1_{33} = \partial_2((2\lambda - r)\sin^2\vartheta) = 2(2\lambda - r)\sin\vartheta\cos\vartheta$$

$$\partial_2 \Gamma^2_{33} = \partial_2(-\sin\vartheta\cos\vartheta) = \sin^2\vartheta - \cos^2\vartheta$$

$$\partial_2 \Gamma^0_{10} = \partial_2\left(\frac{\lambda}{r^2 - 2\lambda r}\right) = 0$$

$$\partial_2 \Gamma^2_{12} = \partial_2\left(\frac{1}{r}\right) = 0$$

$$\partial_2 \Gamma^3_{13} = \partial_2\left(\frac{1}{r}\right) = 0$$

$$\partial_2 \Gamma^3_{23} = \partial_2\left(\frac{\cos\vartheta}{\sin\vartheta}\right) = -\frac{1}{\sin^2\vartheta}.$$

Now we calculate the components of the curvature tensor which do not vanish immediately:

$$R^0_{101} = -R^0_{110} = -\partial_1\Gamma^1_{01} + \Gamma^0_{0r}\Gamma^r_{11} - \Gamma^0_{1r}\Gamma^r_{10} = -\partial_1\Gamma^1_{01} + \Gamma^0_{01}\Gamma^1_{11} - \Gamma^0_{10}\Gamma^0_{10}$$

$$= \frac{2\lambda^2 - 2\lambda r}{(r^2 - 2\lambda r)^2} + \frac{\lambda}{r^2 - 2\lambda r} \cdot \frac{\lambda}{2\lambda r - r^2} - \frac{\lambda^2}{(r^2 - 2\lambda r)^2} = \frac{-2\lambda r}{(r^2 - 2\lambda r)^2}$$

$$R^0_{202} = -R^0_{220} = \Gamma^0_{0r}\Gamma^r_{22} = \Gamma^0_{01}\Gamma^1_{22} = \frac{\lambda}{r^2 - 2\lambda r} \cdot (2\lambda - r) = \frac{\lambda}{r}$$

$$R^0_{303} = -R^0_{330} = \Gamma^0_{0r}\Gamma^r_{33} = \Gamma^0_{01}\Gamma^1_{33} = \frac{\lambda}{r^2 - 2\lambda r} \cdot (2\lambda - r)\sin^2\vartheta = -\frac{\lambda\sin^2\vartheta}{r}$$

$$R^1_{010} = -R^1_{001} = \partial_1\Gamma^1_{00} + \Gamma^1_{1r}\Gamma^r_{00} - \Gamma^1_{0r}\Gamma^r_{01} = \partial_1\Gamma^1_{00} + \Gamma^1_{11}\Gamma^1_{00} - \Gamma^1_{00}\Gamma^0_{01}$$

$$= \frac{6\lambda - 2r}{r^4} + \frac{\lambda}{2\lambda r - r^2} \cdot \frac{r - 2\lambda}{r^3} - \frac{r - 2\lambda}{r^3} \cdot \frac{\lambda}{r^2 - 2\lambda r} = \frac{6\lambda - 2r}{r^4} - \frac{2\lambda}{r^4} = \frac{-4\lambda - 2r}{r^4}$$

$$R^2_{020} = -R^2_{002} = \Gamma^2_{2r}\Gamma^r_{00} = \Gamma^2_{21}\Gamma^1_{00} = \frac{1}{r} \cdot \frac{r - 2\lambda}{r^3} = \frac{r - 2\lambda}{r^4}$$

$$R^3_{030} = -R^3_{003} = \Gamma^3_{3r}\Gamma^r_{00} = \Gamma^3_{31}\Gamma^1_{00} = \frac{1}{r} \cdot \frac{r - 2\lambda}{r^3} = \frac{r - 2\lambda}{r^4}$$

$$R^1_{212} = -R^1_{221} = \partial_1 \Gamma^1_{22} + \Gamma^1_{1r} \Gamma^r_{22} - \Gamma^1_{2r} \Gamma^r_{21} = \partial_1 \Gamma^1_{22} + \Gamma^1_{11} \Gamma^1_{22} - \Gamma^1_{22} \Gamma^2_{21}$$
$$= -1 + \frac{\lambda}{r(2\lambda - r)} \cdot (2\lambda - r) - (2\lambda - r) \cdot \frac{1}{r} = -1 + \frac{\lambda}{r} - \frac{2\lambda}{r} + 1 = -\frac{\lambda}{r}$$

$$R^2_{121} = -R^2_{112} = -\partial_1 \Gamma^2_{21} + \Gamma^2_{2r} \Gamma^r_{11} - \Gamma^2_{1r} \Gamma^r_{12} = -\partial_1 \Gamma^2_{21} + \Gamma^2_{21} \Gamma^1_{11} - \Gamma^2_{12} \Gamma^2_{12}$$
$$= \frac{1}{r^2} + \frac{1}{r} \cdot \frac{\lambda}{r(2\lambda - r)} - \frac{1}{r^2} = \frac{\lambda}{r^2(2\lambda - r)}$$

$$R^1_{313} = -R^1_{331} = \partial_1 \Gamma^1_{33} + \Gamma^1_{1r} \Gamma^r_{33} - \Gamma^1_{3r} \Gamma^r_{31} = \partial_1 \Gamma^1_{33} + \Gamma^1_{11} \Gamma^1_{33} - \Gamma^1_{33} \Gamma^3_{31}$$
$$= -\sin^2 \vartheta + \frac{\lambda}{r(2\lambda - r)} \cdot (2\lambda - r) \sin^2 \vartheta - (2\lambda - r) \sin^2 \vartheta \cdot \frac{1}{r}$$
$$= -\sin^2 \vartheta + \frac{\lambda \sin^2 \vartheta}{r} - \frac{2\lambda \sin^2 \vartheta}{r} + \sin^2 \vartheta = -\frac{\lambda \sin^2 \vartheta}{r}$$

$$R^3_{131} = -R^3_{113} = -\partial_1 \Gamma^3_{31} + \Gamma^3_{3r} \Gamma^r_{11} - \Gamma^3_{1r} \Gamma^r_{13} = -\partial_1 \Gamma^3_{31} + \Gamma^3_{31} \Gamma^1_{11} - \Gamma^3_{13} \Gamma^3_{13}$$
$$= \frac{1}{r^2} + \frac{1}{r} \cdot \frac{\lambda}{r(2\lambda - r)} - \frac{1}{r^2} = \frac{\lambda}{r^2(2\lambda - r)}$$

$$R^2_{323} = -R^2_{332} = \partial_2 \Gamma^2_{33} + \Gamma^2_{2r} \Gamma^r_{33} - \Gamma^2_{3r} \Gamma^r_{32} = \partial_2 \Gamma^2_{33} + \Gamma^2_{21} \Gamma^1_{33} - \Gamma^2_{33} \Gamma^3_{32}$$
$$= \sin^2 \vartheta - \cos^2 \vartheta + \frac{1}{r} \cdot (2\lambda - r) \sin^2 \vartheta + \sin \vartheta \cos \vartheta \cdot \frac{\cos \vartheta}{\sin \vartheta} = \frac{2\lambda \sin^2 \vartheta}{r}$$

$$R^3_{232} = -R^3_{223} = -\partial_2 \Gamma^3_{32} + \Gamma^3_{3r} \Gamma^r_{22} - \Gamma^3_{2r} \Gamma^r_{23} = -\partial_2 \Gamma^3_{32} + \Gamma^3_{31} \Gamma^1_{22} - \Gamma^3_{23} \Gamma^3_{23}$$
$$= \frac{1}{\sin^2 \vartheta} + \frac{1}{r} \cdot (2\lambda - r) - \frac{\cos^2 \vartheta}{\sin^2 \vartheta} = \frac{2\lambda}{r}$$

$$R^3_{312} = -R^3_{321} = \Gamma^3_{13} \Gamma^3_{23} - \Gamma^3_{23} \Gamma^3_{31} = 0$$

$$R^3_{132} = -R^3_{123} = \Gamma^3_{32} \Gamma^2_{21} - \Gamma^3_{23} \Gamma^3_{13} = \frac{\cos \vartheta}{\sin \vartheta} \cdot \frac{1}{r} - \frac{\cos \vartheta}{\sin \vartheta} \cdot \frac{1}{r} = 0$$

$$R^3_{231} = -R^3_{213} = \Gamma^3_{32} \Gamma^2_{12} - \Gamma^3_{13} \Gamma^3_{23} = \frac{\cos \vartheta}{\sin \vartheta} \cdot \frac{1}{r} - \frac{1}{r} \cdot \frac{\cos \vartheta}{\sin \vartheta} = 0$$

$$R^1_{332} = -R^1_{323} = -\partial_2 \Gamma^1_{33} + \Gamma^1_{33} \Gamma^3_{23} - \Gamma^1_{22} \Gamma^2_{33}$$
$$= -2(2\lambda - r) \sin \vartheta \cos \vartheta + (2\lambda - r) \sin^2 \vartheta \cdot \frac{\cos \vartheta}{\sin \vartheta} + (2\lambda - r) \sin \vartheta \cos \vartheta = 0$$

$$R^2_{331} = -R^2_{313} = \Gamma^2_{33} \Gamma^3_{13} - \Gamma^2_{12} \Gamma^2_{33} = -\sin \vartheta \cos \vartheta \cdot \frac{1}{r} - \frac{1}{r} \cdot (-\sin \vartheta \cos \vartheta) = 0.$$

A.2 The Kerr spacetime

In Appendix A.2 we are going to discuss the calculations for the Kerr solution more closely.

The inverse of $(\tilde{g}_{ij})$ A.2.1 First we have to calculate the inverse of $(\tilde{g}_{ij})$. For a matrix in "Kerr"-form we get

$$\begin{pmatrix} a & 0 & 0 & b \\ 0 & c & 0 & 0 \\ 0 & 0 & d & 0 \\ b & 0 & 0 & e \end{pmatrix}^{-1} = \begin{pmatrix} \frac{e}{ae-b^2} & 0 & 0 & \frac{-b}{ae-b^2} \\ 0 & \frac{1}{c} & 0 & 0 \\ 0 & 0 & \frac{1}{d} & 0 \\ \frac{-b}{ae-b^2} & 0 & 0 & \frac{a}{ae-b^2} \end{pmatrix}.$$

A.2. The Kerr spacetime

In case of

$$(\tilde{g}_{ij}) = \begin{pmatrix} -\frac{1}{\lambda}(1 - \frac{2mr}{\rho}) & 0 & 0 & \frac{-2mra\sin^2\vartheta}{\sqrt{\lambda}\rho} \\ 0 & \frac{\rho}{\Delta} & 0 & 0 \\ 0 & 0 & \rho & 0 \\ \frac{-2mra\sin^2\vartheta}{\sqrt{\lambda}\rho} & 0 & 0 & \left(r^2 + a^2 + \frac{2mra^2\sin^2\vartheta}{\rho}\right)\sin^2\vartheta \end{pmatrix},$$

with $\rho := r^2 + a^2\cos^2\vartheta$ and $\Delta := r^2 - 2mr + a^2$, this exactly means that we have to calculate the following:

$$\begin{aligned}
ae - b^2 &= -\frac{1}{\lambda}\left(1 - \frac{2mr}{\rho}\right) \cdot \sin^2\vartheta \left(r^2 + a^2 + \frac{2mra^2\sin^2\vartheta}{\rho}\right) - \frac{4m^2r^2a^2\sin^4\vartheta}{\lambda\rho^2} \\
&= \frac{-\rho + 2mr}{\lambda\rho} \cdot \sin^2\vartheta(r^2 + a^2) + \frac{-\rho + 2mr}{\lambda\rho} \cdot \frac{2mra^2\sin^4\vartheta}{\rho} - \frac{4m^2r^2a^2\sin^4\vartheta}{\lambda\rho^2} \\
&= \frac{(-\rho + 2mr) \cdot \sin^2\vartheta(r^2 + a^2)}{\lambda\rho} - \frac{2mra^2\sin^4\vartheta}{\lambda\rho} + \frac{4m^2r^2a^2\sin^4\vartheta}{\lambda\rho^2} - \frac{4m^2r^2a^2\sin^4\vartheta}{\lambda\rho^2} \\
&= \frac{-\rho\sin^2\vartheta(a^2 + r^2) + 2mr^3\sin^2\vartheta + 2mra^2\sin^2\vartheta\cos^2\vartheta}{\lambda\rho} \\
&= \frac{-\rho\sin^2\vartheta(a^2 + r^2) + 2mr\sin^2\vartheta(r^2 + a^2\cos^2\vartheta)}{\lambda\rho} \\
&= \frac{\rho(-\sin^2\vartheta(r^2 + a^2 - 2mr))}{\lambda\rho} = -\frac{1}{\lambda}\sin^2\vartheta \cdot \Delta
\end{aligned}$$

$$\frac{b}{ae - b^2} = \frac{\frac{2mra\sin^2\vartheta}{\sqrt{\lambda}\rho}}{-\frac{1}{\lambda}\sin^2\vartheta \cdot \Delta} = -\sqrt{\lambda}\frac{2mra}{\rho\Delta}$$

$$\frac{a}{ae - b^2} = \frac{-\frac{1}{\lambda}\left(a - \frac{2mr}{\rho}\right)}{-\frac{1}{\lambda}\sin^2\vartheta \cdot \Delta} = \frac{\rho - 2mr}{\sin^2\vartheta \cdot \rho\Delta}$$

$$\begin{aligned}
\frac{e}{ae - b^2} &= \frac{\left(r^2 + a^2 + \frac{2mra^2\sin^2\vartheta}{\rho}\right)\sin^2\vartheta}{-\frac{1}{\lambda}\sin^2\vartheta \cdot \Delta} = \frac{-\lambda((r^2 + a^2)\rho + 2mra^2 - 2mra^2\cos^2\vartheta)}{\Delta\rho} \\
&= \frac{-\lambda}{\Delta\rho}(r^4 + r^2a^2\cos^2\vartheta + a^2r^2 + a^4\cos^2\vartheta + 2mra^2 - 2mra^2\cos^2\vartheta) \\
&= \frac{-\lambda}{\Delta\rho}(a^2\cos^2\vartheta(r^2 - 2mr + a^2) + r^2(r^2 - 2mr + a^2) + 2mr(r^2 + a^2)) \\
&= \frac{-\lambda}{\Delta\rho}(\Delta\rho + 2mr(r^2 + a^2)) = -\lambda\left(1 + \frac{2mr(r^2 + a^2)}{\Delta\rho}\right).
\end{aligned}$$

If we now again replace m by λ we get for $(h^{ij}(\lambda))$:

$$(h^{ij}(\lambda)) = \begin{pmatrix} -\lambda\left(1 + \frac{2mr(r^2+a^2)}{\Delta\rho}\right) & 0 & 0 & -\sqrt{\lambda}\frac{2mra}{\rho\Delta} \\ 0 & \frac{\tilde{\Delta}}{\rho} & 0 & 0 \\ 0 & 0 & \frac{1}{\rho} & 0 \\ -\sqrt{\lambda}\frac{2mra}{\rho\Delta} & 0 & 0 & \frac{\rho-2mr}{\sin^2\vartheta\cdot\rho\Delta} \end{pmatrix},$$

with $\rho := r^2 + a^2\cos^2\vartheta$ and $\tilde{\Delta} := r^2 - 2\lambda r + a^2$.

A.2. The Kerr spacetime

The Christoffel symbols A.2.2 For the Christoffel symbols of the extended Kerr spacetime we first have to invert $(g_{ij}(\lambda))$. The calculations are very similar to the ones above:

$$\begin{pmatrix} a & 0 & 0 & b \\ 0 & c & 0 & 0 \\ 0 & 0 & d & 0 \\ b & 0 & 0 & e \end{pmatrix}^{-1} = \begin{pmatrix} \frac{e}{ae-b^2} & 0 & 0 & \frac{-b}{ae-b^2} \\ 0 & \frac{1}{c} & 0 & 0 \\ 0 & 0 & \frac{1}{d} & 0 \\ \frac{-b}{ae-b^2} & 0 & 0 & \frac{a}{ae-b^2} \end{pmatrix}.$$

Now we have to consider

$$(g_{ij}(\lambda)) = \begin{pmatrix} 1 - \frac{2\lambda r}{\rho} & 0 & 0 & \frac{2\lambda^{\frac{3}{2}} r a \sin^2 \vartheta}{\rho} \\ 0 & -\lambda \frac{\rho}{\widetilde{\Delta}} & 0 & 0 \\ 0 & 0 & -\lambda \rho & 0 \\ \frac{2\lambda^{\frac{3}{2}} r a \sin^2 \vartheta}{\rho} & 0 & 0 & -\lambda \left(r^2 + a^2 + \frac{2\lambda r a^2 \sin^2 \vartheta}{\rho} \right) \sin^2 \vartheta \end{pmatrix}$$

with $\rho := r^2 + a^2 \cos^2 \vartheta$ and $\widetilde{\Delta} := r^2 - 2\lambda r + a^2$.
So we have to calculate the following:

$$ae - b^2 = \left(1 - \frac{2\lambda r}{\rho}\right) \cdot \left(-\lambda \sin^2 \vartheta \left(r^2 + a^2 + \frac{2\lambda r a^2 \sin^2 \vartheta}{\rho}\right)\right) - \frac{4\lambda^3 r^2 a^2 \sin^4 \vartheta}{\rho^2}$$

$$= -\lambda \sin^2 \vartheta (r^2 + a^2) - \frac{2\lambda^2 r a^2 \sin^4 \vartheta}{\rho} + \frac{2\lambda^2 r \sin^2 \vartheta (r^2 + a^2)}{\rho} + \frac{4\lambda^3 r^2 a^2 \sin^4 \vartheta}{\rho^2}$$

$$- \frac{4\lambda^3 r^2 a^2 \sin^4 \vartheta}{\rho^2}$$

$$= -\lambda \sin^2 \vartheta \left(\frac{(r^2 + a^2)(r^2 + a^2 \cos^2 \vartheta) - 2\lambda r a^2 \cos^2 \vartheta - 2\lambda r^3}{\rho} \right)$$

$$= -\lambda \sin^2 \vartheta \left(\frac{(r^2 + a^2)(r^2 + a^2 \cos^2 \vartheta) - 2\lambda r (a^2 \cos^2 \vartheta + r^2)}{\rho} \right)$$

$$= -\lambda \sin^2 \vartheta \left(\frac{\rho (r^2 - 2\lambda r + a^2)}{\rho} \right) = -\lambda \sin^2 \vartheta \cdot \widetilde{\Delta}$$

$$\frac{e}{ae - b^2} = \frac{-\lambda \sin^2 \vartheta \left(r^2 + a^2 + \frac{2\lambda r a^2 \sin^2 \vartheta}{\rho} \right)}{-\lambda \sin^2 \vartheta \cdot \widetilde{\Delta}} = \frac{(r^2 + a^2)\rho + 2\lambda r a^2 - 2\lambda r a^2 \cos^2 \vartheta}{\widetilde{\Delta} \rho}$$

$$= \frac{(r^4 + r^2 a^2 \cos^2 \vartheta + a^2 r^2 + a^4 \cos^2 \vartheta + 2\lambda r a^2 - 2\lambda r a^2 \cos^2 \vartheta)}{\widetilde{\Delta} \rho}$$

$$= \frac{1}{\widetilde{\Delta} \rho} (a^2 \cos^2 \vartheta (r^2 - 2\lambda r + a^2) + r^2 (r^2 - 2\lambda r + a^2) + 2\lambda r (r^2 + a^2))$$

$$= \frac{(\widetilde{\Delta} \rho + 2\lambda r (r^2 + a^2))}{\widetilde{\Delta} \rho} = 1 + \frac{2\lambda r (r^2 + a^2)}{\widetilde{\Delta} \rho}$$

$$-\frac{b}{ae - b^2} = \frac{-2\lambda^{\frac{3}{2}} r a \sin^2 \vartheta}{-\lambda \sin^2 \vartheta \cdot \widetilde{\Delta} \rho} = \frac{2\sqrt{\lambda} r a}{\widetilde{\Delta} \rho}$$

$$\frac{a}{ae - b^2} = \frac{1 - \frac{2\lambda r}{\rho}}{-\lambda \sin^2 \vartheta \cdot \widetilde{\Delta}} = \frac{\rho - 2\lambda r}{-\lambda \sin^2 \vartheta \cdot \widetilde{\Delta} \rho}.$$

A.2. The Kerr spacetime

Now we calculate the derivations of the components of $(g_{ij}(\lambda))$ with regard to the corresponding coordinates. As the components only depend on r and ϑ, we only have to calculate the following:

$$\frac{\partial a}{\partial r} = \partial_r \left(1 - \frac{2\lambda r}{r^2 + a^2 \cos^2 \vartheta}\right) = \frac{-2\lambda(r^2 + a^2 \cos^2 \vartheta) + 2\lambda r \cdot 2r}{\rho^2} = \frac{2\lambda(r^2 - a^2 \cos^2 \vartheta)}{\rho^2}$$

$$\frac{\partial b}{\partial r} = \partial_r \left(\frac{2\lambda^{\frac{3}{2}} r a \sin^2 \vartheta}{r^2 + a^2 \cos^2 \vartheta}\right) = \frac{2\lambda^{\frac{3}{2}} a \sin^2 \vartheta (r^2 + a^2 \cos^2 \vartheta) - 4\lambda^{\frac{3}{2}} r^2 a \sin^2 \vartheta}{\rho^2}$$

$$= \frac{-2\lambda^{\frac{3}{2}} a \sin^2 \vartheta (r^2 - a^2 \cos^2 \vartheta)}{\rho^2}$$

$$\frac{\partial c}{\partial r} = \partial_r \left(-\lambda \frac{r^2 + a^2 \cos^2 \vartheta}{r^2 - 2\lambda r + a^2}\right) = -\lambda \frac{2r(r^2 - 2\lambda r + a^2) - (r^2 + a^2 \cos^2 \vartheta)(2r - 2\lambda)}{\widetilde{\Delta}^2}$$

$$= -\lambda \frac{2ra^2 \sin^2 \vartheta - 2\lambda(r^2 - a^2 \cos^2 \vartheta)}{\widetilde{\Delta}^2}$$

$$\frac{\partial d}{\partial r} = \partial_r \left(-\lambda(r^2 + a^2 \cos^2 \vartheta)\right) = -2\lambda r$$

$$\frac{\partial e}{\partial r} = \partial_r \left(-\lambda \sin^2 \vartheta \left(r^2 + a^2 + \frac{2\lambda r a^2 \sin^2 \vartheta}{r^2 + a^2 \cos^2 \vartheta}\right)\right)$$

$$= -2\lambda r \sin^2 \vartheta - \lambda \sin^2 \vartheta \frac{2\lambda a^2 \sin^2 \vartheta (r^2 + a^2 \cos^2 \vartheta) - 4\lambda r^2 a^2 \sin^2 \vartheta}{\rho^2}$$

$$= -2\lambda \sin^2 \vartheta \left(\frac{r\rho^2 - \lambda a^2 \sin^2 \vartheta (r^2 - a^2 \cos^2 \vartheta)}{\rho^2}\right)$$

$$\frac{\partial a}{\partial \vartheta} = \partial_\vartheta \left(1 - \frac{2\lambda r}{r^2 + a^2 \cos^2 \vartheta}\right) = \frac{-2\lambda r \cdot 2a^2 \cos \vartheta \sin \vartheta}{\rho^2} = \frac{-4\lambda r a^2 \sin \vartheta \cos \vartheta}{\rho^2}$$

$$\frac{\partial b}{\partial \vartheta} = \partial_\vartheta \left(\frac{2\lambda^{\frac{3}{2}} r a \sin^2 \vartheta}{r^2 + a^2 \cos^2 \vartheta}\right) = \frac{4\lambda^{\frac{3}{2}} r a \sin \vartheta \cos \vartheta (r^2 + a^2 \cos^2 \vartheta) + 4\lambda^{\frac{3}{2}} r a^3 \sin^3 \vartheta \cos \vartheta}{\rho^2}$$

$$= \frac{4\lambda^{\frac{3}{2}} r a \sin \vartheta \cos \vartheta (r^2 + a^2 \sin^2 \vartheta + a^2 \cos^2 \vartheta)}{\rho^2} = \frac{4\lambda^{\frac{3}{2}} r a \sin \vartheta \cos \vartheta (r^2 + a^2)}{\rho^2}$$

$$\frac{\partial c}{\partial \vartheta} = \partial_\vartheta \left(-\lambda \frac{r^2 + a^2 \cos^2 \vartheta}{r^2 - 2\lambda r + a^2}\right) = \frac{2\lambda a^2 \sin \vartheta \cos \vartheta}{\widetilde{\Delta}}$$

$$\frac{\partial d}{\partial \vartheta} = \partial_\vartheta \left(-\lambda(r^2 + a^2 \cos^2 \vartheta)\right) = 2\lambda a^2 \sin \vartheta \cos \vartheta$$

$$\frac{\partial e}{\partial \vartheta} = \partial_\vartheta \left(-\lambda(r^2 + a^2)\sin^2 \vartheta - \frac{2\lambda^2 r a^2 \sin^4 \vartheta}{r^2 + a^2 \cos^2 \vartheta}\right) = -2\lambda(r^2 + a^2)\sin \vartheta \cos \vartheta$$

$$- \frac{8\lambda^2 r a^2 \sin^3 \vartheta \cos \vartheta (r^2 + a^2 \cos^2 \vartheta) + 2\lambda^2 r a^2 \sin^4 \vartheta (2a^2 \sin \vartheta \cos \vartheta)}{\rho^2}$$

$$= \frac{-2\lambda \sin \vartheta \cos \vartheta}{\rho^2} \left(r^6 + a^2 r^4 + 2r^4 a^2 \cos^2 \vartheta + 2r^2 a^4 \cos^2 \vartheta + r^2 a^4 \cos^4 \vartheta + a^6 \cos^4 \vartheta\right.$$
$$\left. + 4\lambda r^3 a^2 - 4\lambda r^3 a^2 \cos^2 \vartheta + 2\lambda r a^4 - 2\lambda r a^4 \cos^2 \vartheta + 2\lambda r a^4 \cos^2 \vartheta - 2\lambda r a^4 \cos^4 \vartheta\right)$$

$$= \frac{-2\lambda \sin \vartheta \cos \vartheta}{\rho^2} \left(a^4 \cos^4 \vartheta (a^2 - 2\lambda r + r^2) + 2r^2 a^2 \cos^2 \vartheta (a^2 - 2\lambda r + r^2)\right.$$
$$\left. + r^4(a^2 - 2\lambda r + r^2) + 2\lambda r^5 + 4\lambda r^3 a^2 + 2\lambda r a^4\right)$$

$$= \frac{-2\lambda \sin \vartheta \cos \vartheta (\widetilde{\Delta} \rho^2 + 2\lambda r (r^2 + a^2)^2)}{\rho^2}.$$

130

A.2. The Kerr spacetime

Now we are able to calculate the Christoffel symbols of the extended Kerr spacetime. We get:

$$\Gamma^0_{00} = 0$$

$$\Gamma^1_{00} = \frac{1}{2}g^{11}(\partial_0 g_{01} + \partial_0 g_{01} - \partial_1 g_{00}) = -\frac{1}{2}\left(-\frac{1}{\lambda} \cdot \frac{\tilde{\Delta}}{\rho} \cdot \frac{2\lambda(r^2 - a^2\cos^2\vartheta)}{\rho^2}\right)$$

$$= \frac{\tilde{\Delta}(r^2 - a^2\cos^2\vartheta)}{\rho^3}$$

$$\Gamma^2_{00} = \frac{1}{2}g^{22}(-\partial_2 g_{00}) = \frac{1}{2}\left(-\frac{1}{\lambda\rho} \cdot \frac{4\lambda r a^2 \sin\vartheta \cos\vartheta}{\rho^2}\right) = \frac{-2ra^2 \sin\vartheta \cos\vartheta}{\rho^3}$$

$$\Gamma^3_{00} = 0$$

$$\Gamma^0_{01} = \Gamma^0_{10} = \frac{1}{2}g^{00}(\partial_0 g_{10} + \partial_1 g_{00} - \partial_0 g_{01}) + \frac{1}{2}g^{03}(\partial_0 g_{13} + \partial_1 g_{03} - \partial_3 g_{01})$$

$$= \frac{1}{2}\left(\frac{(\tilde{\Delta}\rho + 2\lambda r(r^2 + a^2))}{\tilde{\Delta}\rho} \cdot \frac{2\lambda(r^2 - a^2\cos^2\vartheta)}{\rho^2}\right.$$

$$\left. + \frac{2\sqrt{\lambda}ra}{\tilde{\Delta}\rho} \cdot \frac{-2\lambda^{\frac{3}{2}}a\sin^2\vartheta(r^2 - a^2\cos^2\vartheta)}{\rho^2}\right)$$

$$= \frac{\lambda(r^2 - a^2\cos^2\vartheta)}{\rho^3\tilde{\Delta}}(\rho\tilde{\Delta} + 2\lambda r((r^2 + a^2) - a^2\sin^2\vartheta))$$

$$= \frac{\lambda(r^2 - a^2\cos^2\vartheta)}{\rho^3\tilde{\Delta}}(\rho\tilde{\Delta} + 2\lambda r \cdot \rho)$$

$$= \frac{\lambda(r^2 - a^2\cos^2\vartheta)}{\rho^3\tilde{\Delta}}(\rho(r^2 - 2\lambda r + a^2 + 2\lambda r))$$

$$= \frac{\lambda(r^2 - a^2\cos^2\vartheta)(r^2 + a^2)}{\rho^2\tilde{\Delta}}$$

$$\Gamma^0_{02} = \Gamma^0_{20} = \frac{1}{2}g^{00}(\partial_0 g_{20} + \partial_2 g_{00} - \partial_0 g_{02}) + \frac{1}{2}g^{03}(\partial_0 g_{23} + \partial_2 g_{03} - \partial_3 g_{02})$$

$$= \frac{1}{2}\left(\frac{(\tilde{\Delta}\rho + 2\lambda r(r^2 + a^2))}{\tilde{\Delta}\rho} \cdot \frac{-4\lambda ra^2 \sin\vartheta\cos\vartheta}{\rho^2}\right.$$

$$\left. + \frac{2\sqrt{\lambda}ra}{\tilde{\Delta}\rho} \cdot \frac{4\lambda^{\frac{3}{2}}ra\sin\vartheta\cos\vartheta(r^2 + a^2)}{\rho^2}\right)$$

$$= \frac{4\lambda ra^2 \sin\vartheta\cos\vartheta}{\rho^3\tilde{\Delta}}(-\rho\tilde{\Delta} - 2\lambda r(r^2 + a^2) + 2\lambda r(r^2 + a^2))$$

$$= \frac{-4\lambda ra^2 \sin\vartheta\cos\vartheta}{\rho^2}$$

$$\Gamma^0_{03} = \Gamma^0_{30} = 0$$

$$\Gamma^0_{11} = 0$$

$$\Gamma^0_{22} = 0$$

$$\Gamma^0_{33} = 0$$

$$\Gamma^1_{01} = \Gamma^1_{10} = 0$$

$$\Gamma^2_{02} = \Gamma^2_{20} = 0$$

$\Gamma^3_{03} = \Gamma^3_{30} = 0$

$\Gamma^2_{01} = \Gamma^2_{10} = 0$

$\Gamma^1_{02} = \Gamma^1_{20} = 0$

$\Gamma^0_{12} = \Gamma^0_{21} = 0$

$$\Gamma^3_{01} = \Gamma^3_{10} = \frac{1}{2}g^{30}(\partial_0 g_{10} + \partial_1 g_{00} - \partial_0 g_{01}) + \frac{1}{2}g^{33}(\partial_0 g_{13} + \partial_1 g_{03} - \partial_3 g_{01})$$

$$= \frac{1}{2}\left(\frac{2\sqrt{\lambda}ra}{\widetilde{\Delta}\rho} \cdot \frac{2\lambda(r^2 - a^2\cos^2\vartheta)}{\rho^2}\right.$$

$$\left. + \frac{\rho - 2\lambda r}{-\lambda\sin^2\vartheta \cdot \widetilde{\Delta}\rho} \cdot \frac{-2\lambda^{\frac{3}{2}}a\sin^2\vartheta(r^2 - a^2\cos^2\vartheta)}{\rho^2}\right)$$

$$= \frac{\sqrt{\lambda}a(r^2 - a^2\cos^2\vartheta)(2\lambda r + \rho - 2\lambda r)}{\widetilde{\Delta}\rho^3} = \frac{\sqrt{\lambda}a(r^2 - a^2\cos^2\vartheta)}{\widetilde{\Delta}\rho^2}$$

$$\Gamma^3_{02} = \Gamma^3_{20} = \frac{1}{2}g^{30}(\partial_0 g_{20} + \partial_2 g_{00} - \partial_0 g_{02}) + \frac{1}{2}g^{33}(\partial_0 g_{23} + \partial_2 g_{03} - \partial_3 g_{02})$$

$$= \frac{1}{2}\left(\frac{2\sqrt{\lambda}ra}{\widetilde{\Delta}\rho} \cdot \frac{-4\lambda ra^2\sin\vartheta\cos\vartheta}{\rho^2}\right.$$

$$\left. + \frac{\rho - 2\lambda r}{-\lambda\sin^2\vartheta \cdot \widetilde{\Delta}\rho} \cdot \frac{4\lambda^{\frac{3}{2}}ra\sin\vartheta\cos\vartheta(r^2 + a^2)}{\rho^2}\right)$$

$$= \frac{-2\sqrt{\lambda}ra\cos\vartheta}{\sin\vartheta \cdot \widetilde{\Delta}\rho^3}(\rho(r^2 + a^2) - 2\lambda r(r^2 + a^2 - a^2\sin^2\vartheta))$$

$$= \frac{-2\sqrt{\lambda}ra\cos\vartheta}{\sin\vartheta \cdot \widetilde{\Delta}\rho^3}(\rho(r^2 + a^2) - 2\lambda r \cdot \rho) = \frac{-2\sqrt{\lambda}ra\cos\vartheta}{\sin\vartheta \cdot \rho^2}$$

$$\Gamma^0_{13} = \Gamma^0_{31} = \frac{1}{2}g^{00}(\partial_1 g_{30} + \partial_3 g_{10} - \partial_0 g_{13}) + \frac{1}{2}g^{03}(\partial_1 g_{33} + \partial_3 g_{13} - \partial_3 g_{13})$$

$$= \frac{1}{2}\left(\frac{\rho\widetilde{\Delta} + 2\lambda r(r^2 + a^2)}{\rho\widetilde{\Delta}} \cdot \frac{-2\lambda^{\frac{3}{2}}a\sin^2\vartheta(r^2 - a^2\cos^2\vartheta)}{\rho^2}\right.$$

$$\left. + \frac{2\sqrt{\lambda}ra}{\widetilde{\Delta}\rho} \cdot \frac{-2\lambda\sin^2\vartheta}{\rho^2}(r\rho^2 - \lambda a^2\sin^2\vartheta(r^2 - a^2\cos^2\vartheta))\right)$$

$$= \frac{-\lambda^{\frac{3}{2}}a\sin^2\vartheta}{\rho^3\widetilde{\Delta}}[(\rho\widetilde{\Delta} + 2\lambda r(r^2 + a^2)) \cdot (r^2 - a^2\cos^2\vartheta) + 2r^2\rho^2$$

$$- 2\lambda ra^2\sin^2\vartheta(r^2 - a^2\cos^2\vartheta)]$$

$$= \frac{-\lambda^{\frac{3}{2}}a\sin^2\vartheta}{\rho^3\widetilde{\Delta}}[(\rho\widetilde{\Delta} + 2\lambda r(r^2 + a^2 - a^2\sin^2\vartheta)) \cdot (r^2 - a^2\cos^2\vartheta) + 2r^2\rho^2]$$

$$= \frac{-\lambda^{\frac{3}{2}}a\sin^2\vartheta}{\rho^3\widetilde{\Delta}}[\rho(r^2 + a^2) \cdot (r^2 - a^2\cos^2\vartheta) + 2r^2\rho^2]$$

$$= \frac{-\lambda^{\frac{3}{2}}a\sin^2\vartheta}{\rho^2\widetilde{\Delta}}[(r^2 + a^2) \cdot (r^2 - a^2\cos^2\vartheta) + 2r^2\rho]$$

$$\Gamma^0_{23} = \Gamma^0_{32} = \frac{1}{2}g^{00}(\partial_2 g_{30} + \partial_3 g_{20} - \partial_0 g_{23}) + \frac{1}{2}g^{03}(\partial_2 g_{33} + \partial_3 g_{23} - \partial_3 g_{23})$$

A.2. The Kerr spacetime

$$= \frac{1}{2}\left(\frac{\rho\widetilde{\Delta} + 2\lambda r(r^2+a^2)}{\rho\widetilde{\Delta}} \cdot \frac{4\lambda^{\frac{3}{2}}ra\sin\vartheta\cos\vartheta(r^2+a^2)}{\rho^2}\right.$$
$$\left.+\frac{2\sqrt{\lambda}ra}{\widetilde{\Delta}} \cdot \frac{-2\lambda\sin\vartheta\cos\vartheta}{\rho^2}(\widetilde{\Delta}\rho^2 + 2\lambda r(r^2+a^2)^2)\right)$$
$$= \frac{2\lambda^{\frac{3}{2}}ra\sin\vartheta\cos\vartheta}{\rho^3\widetilde{\Delta}}[\rho\widetilde{\Delta}((r^2+a^2)-\rho) + 2\lambda r(r^2+a^2)^2 - 2\lambda r(r^2+a^2)^2]$$
$$= \frac{2\lambda^{\frac{3}{2}}ra\sin\vartheta\cos\vartheta}{\rho^2}[(r^2+a^2-r^2-a^2\cos^2\vartheta)] = \frac{2\lambda^{\frac{3}{2}}ra^3\sin^3\vartheta\cos\vartheta}{\rho^2}$$

$$\Gamma^1_{30} = \Gamma^1_{03} = \frac{1}{2}g^{11}(\partial_0 g_{31} + \partial_3 g_{01} - \partial_1 g_{03}) = \frac{1}{2}\left(-\frac{\widetilde{\Delta}}{\lambda\rho}\right) \cdot \frac{2\lambda^{\frac{3}{2}}a\sin^2\vartheta(r^2-a^2\cos^2\vartheta)}{\rho^2}$$
$$= -\frac{\sqrt{\lambda\widetilde{\Delta}}\cdot a\sin^2\vartheta(r^2-a^2\cos^2\vartheta)}{\rho^3}$$

$$\Gamma^2_{30} = \Gamma^2_{03} = \frac{1}{2}g^{22}(\partial_0 g_{32} + \partial_3 g_{02} - \partial_2 g_{03}) = \frac{1}{2}\left(-\frac{1}{\lambda\rho}\right) \cdot \left(-\frac{4\lambda^{\frac{3}{2}}ra\sin\vartheta\cos\vartheta(r^2+a^2)}{\rho^2}\right)$$
$$= \frac{2\sqrt{\lambda}ra\sin\vartheta\cos\vartheta(r^2+a^2)}{\rho^3}$$

$$\Gamma^1_{11} = \frac{1}{2}g^{11}(\partial_1 g_{11} + \partial_1 g_{11} - \partial_1 g_{11})$$
$$= \frac{1}{2}\left(-\frac{\widetilde{\Delta}}{\lambda\rho}\right) \cdot \left(-\lambda\frac{2ra^2\sin^2\vartheta - 2\lambda(r^2-a^2\cos^2\vartheta)}{\widetilde{\Delta}^2}\right)$$
$$= \frac{ra^2\sin^2\vartheta - \lambda(r^2-a^2\cos^2\vartheta)}{\rho\widetilde{\Delta}}$$

$$\Gamma^2_{11} = \frac{1}{2}g^{22}(\partial_1 g_{12} + \partial_1 g_{12} - \partial_2 g_{11}) = \frac{1}{2}\left(-\frac{1}{\lambda\rho}\right) \cdot \left(-\frac{2\lambda a^2\sin\vartheta\cos\vartheta}{\widetilde{\Delta}}\right) = \frac{a^2\sin\vartheta\cos\vartheta}{\rho\widetilde{\Delta}}$$

$$\Gamma^3_{11} = 0$$

$$\Gamma^1_{21} = \Gamma^1_{12} = \frac{1}{2}g^{11}(\partial_2 g_{11} + \partial_1 g_{21} - \partial_1 g_{21}) = \frac{1}{2}\left(-\frac{\widetilde{\Delta}}{\lambda\rho}\right) \cdot \left(\frac{2\lambda a^2\sin\vartheta\cos\vartheta}{\widetilde{\Delta}}\right) = \frac{-a^2\sin\vartheta\cos\vartheta}{\rho}$$

$$\Gamma^1_{31} = \Gamma^1_{13} = 0$$
$$\Gamma^3_{12} = \Gamma^3_{21} = 0$$
$$\Gamma^2_{13} = \Gamma^2_{31} = 0$$
$$\Gamma^1_{23} = \Gamma^1_{32} = 0$$

$$\Gamma^1_{22} = \frac{1}{2}g^{11}(\partial_2 g_{21} + \partial_2 g_{21} - \partial_1 g_{22}) = \frac{1}{2}\left(\frac{-\widetilde{\Delta}}{\lambda\rho}\right) \cdot (2\lambda r) = -\frac{\widetilde{\Delta}r}{\rho}$$

$$\Gamma^3_{22} = 0$$

$$\Gamma^2_{21} = \Gamma^2_{12} = \frac{1}{2}g^{22}(\partial_2 g_{12} + \partial_1 g_{22} - \partial_2 g_{21}) = \frac{1}{2}\left(\frac{-1}{\lambda\rho}\right) \cdot (-2\lambda r) = \frac{r}{\rho}$$

$$\Gamma^2_{22} = \frac{1}{2}g^{22}(\partial_2 g_{22} + \partial_2 g_{22} - \partial_2 g_{22}) = \frac{1}{2}\left(-\frac{1}{\lambda\rho}\right) \cdot 2\lambda a^2\cos\vartheta\sin\vartheta = -\frac{a^2\cos\vartheta\sin\vartheta}{\rho}$$

$$\Gamma^2_{23} = \Gamma^2_{32} = 0$$

A.2. The Kerr spacetime

$$\Gamma^1_{33} = \frac{1}{2}g^{11}(\partial_3 g_{31} + \partial_3 g_{31} - \partial_1 g_{33}) = \frac{1}{2}\left(-\frac{\widetilde{\Delta}}{\lambda\rho}\right) \cdot \frac{2\lambda\sin^2\vartheta(r\rho^2 - \lambda a^2\sin^2\vartheta(r^2 - a^2\cos\vartheta))}{\rho^2}$$

$$= -\frac{\widetilde{\Delta}\sin^2\vartheta(r\rho^2 - \lambda a^2\sin^2\vartheta(r^2 - a^2\cos\vartheta))}{\rho^3}$$

$$\Gamma^2_{33} = \frac{1}{2}g^{22}(\partial_3 g_{32} + \partial_3 g_{32} - \partial_2 g_{33}) = \frac{1}{2}\left(-\frac{1}{\lambda\rho}\right) \cdot \frac{2\lambda\sin\vartheta\cos\vartheta(\widetilde{\Delta}\rho^2 + 2\lambda r(r^2 + a^2)^2)}{\rho^2}$$

$$= -\frac{\sin\vartheta\cos\vartheta(\widetilde{\Delta}\rho^2 + 2\lambda r(r^2 + a^2)^2)}{\rho^3}$$

$$\Gamma^3_{33} = 0$$

$$\Gamma^3_{31} = \Gamma^3_{13} = \frac{1}{2}g^{30}(\partial_3 g_{10} + \partial_1 g_{30} - \partial_0 g_{31}) + \frac{1}{2}g^{33}(\partial_3 g_{13} + \partial_1 g_{33} - \partial_3 g_{31})$$

$$= \frac{1}{2}\left(\frac{2\sqrt{\lambda}ra}{\widetilde{\Delta}\rho} \cdot \left(-\frac{2\lambda^{\frac{3}{2}}a\sin^2\vartheta(r^2 - a^2\cos^2\vartheta)}{\rho^2}\right)\right)$$

$$+ \frac{1}{2} \cdot \frac{\rho - 2\lambda r}{-\lambda\sin^2\vartheta \cdot \widetilde{\Delta}\rho} \cdot \frac{-2\lambda\sin^2\vartheta(r\rho^2 - \lambda a^2\sin^2\vartheta(r^2 - a^2\cos\vartheta))}{\rho^2}$$

$$= \frac{-2\lambda^2 ra^2\sin^2\vartheta(r^2 - a^2\cos^2\vartheta)}{\widetilde{\Delta}\rho^3} + \frac{(\rho - 2\lambda r)(r\rho^2 - \lambda a^2\sin^2\vartheta(r^2 - a^2\cos\vartheta))}{\widetilde{\Delta}\rho^3}$$

$$= \frac{-\lambda(2r^2\rho - a^2\sin^2\vartheta(r^2 - a^2\cos\vartheta)) + r\rho^2}{\widetilde{\Delta}\rho^2}$$

$$\Gamma^3_{32} = \Gamma^3_{23} = \frac{1}{2}g^{30}(\partial_3 g_{20} + \partial_2 g_{30} - \partial_0 g_{32}) + \frac{1}{2}g^{33}(\partial_3 g_{23} + \partial_2 g_{33} - \partial_3 g_{32})$$

$$= \frac{1}{2}\left(\frac{2\sqrt{\lambda}ra}{\widetilde{\Delta}\rho} \cdot \left(\frac{4\lambda^{\frac{3}{2}}ra\sin\vartheta\cos\vartheta(r^2 + a^2)}{\rho^2}\right)\right)$$

$$+ \frac{1}{2} \cdot \frac{\rho - 2\lambda r}{-\lambda\sin^2\vartheta \cdot \widetilde{\Delta}\rho} \cdot \frac{-2\lambda\cos\vartheta\sin\vartheta(\widetilde{\Delta}\rho^2 + 2\lambda r(a^2 + r^2)^2)}{\rho^2}$$

$$= \frac{\cos\vartheta(4\lambda^2 r^2 a^2\sin^2\vartheta(r^2 + a^2) + (\rho - 2\lambda r)(\widetilde{\Delta}\rho^2 + 2\lambda r(r^2 + a^2)^2))}{\sin\vartheta \cdot \widetilde{\Delta}\rho^3}$$

$$= \frac{\cos\vartheta(4\lambda^2 r^2 a^2\sin^2\vartheta(r^2 + a^2) + \rho^3\widetilde{\Delta} - 2\lambda r\widetilde{\Delta}\rho^2 + 2\lambda r(r^2 + a^2)^2 \cdot \rho)}{\sin\vartheta \cdot \widetilde{\Delta}\rho^3}$$

$$+ \frac{\cos\vartheta(-4\lambda^2 r^2(r^2 + a^2)^2)}{\sin\vartheta \cdot \widetilde{\Delta}\rho^3}$$

$$= \frac{\cos\vartheta(-4\lambda^2 r^2(r^2 + a^2)(-a^2\sin^2\vartheta + r^2 + a^2) + \rho^3\widetilde{\Delta} - 2\lambda r\widetilde{\Delta}\rho^2)}{\sin\vartheta \cdot \widetilde{\Delta}\rho^3}$$

$$+ \frac{\cos\vartheta(2\lambda r(r^2 + a^2)^2 \cdot \rho)}{\sin\vartheta \cdot \widetilde{\Delta}\rho^3}$$

$$= \frac{\cos\vartheta(2\lambda r \cdot \rho(r^2 + a^2)(-2\lambda r + r^2 + a^2) + \rho^3\widetilde{\Delta} - 2\lambda r\widetilde{\Delta}\rho^2)}{\sin\vartheta \cdot \widetilde{\Delta}\rho^3}$$

$$= \frac{\cos\vartheta(2\lambda r \cdot \rho\widetilde{\Delta}(r^2 + a^2 - r^2 - a^2\cos^2\vartheta) + \rho^3\widetilde{\Delta})}{\sin\vartheta \cdot \widetilde{\Delta}\rho^3} = \frac{\cos\vartheta(2\lambda ra^2\sin^2\vartheta + \rho^2)}{\sin\vartheta \cdot \rho^2}.$$

A.2. The Kerr spacetime

The Christoffel symbols of the limit connection A.2.3 The Christoffel symbols of the limit connection have the form:

$$\Gamma^1_{00} = \frac{(r^2 + a^2)(r^2 - a^2 \cos^2 \vartheta)}{(r^2 + a^2 \cos^2 \vartheta)^3}$$

$$\Gamma^2_{00} = \frac{-2ra^2 \sin \vartheta \cos \vartheta}{(r^2 + a^2 \cos^2 \vartheta)^3}$$

$$\Gamma^1_{11} = \frac{ra^2 \sin^2 \vartheta}{(r^2 + a^2 \cos^2 \vartheta)(r^2 + a^2)}$$

$$\Gamma^2_{11} = \frac{a^2 \sin \vartheta \cos \vartheta}{(r^2 + a^2 \cos^2 \vartheta)(r^2 + a^2)}$$

$$\Gamma^1_{21} = \Gamma^1_{12} = \frac{-a^2 \sin \vartheta \cos \vartheta}{r^2 + a^2 \cos^2 \vartheta}$$

$$\Gamma^1_{22} = -\frac{(a^2 + r^2)r}{r^2 + a^2 \cos^2 \vartheta}$$

$$\Gamma^2_{21} = \Gamma^2_{12} = \frac{r}{r^2 + a^2 \cos^2 \vartheta}$$

$$\Gamma^2_{22} = -\frac{a^2 \cos \vartheta \sin \vartheta}{r^2 + a^2 \cos^2 \vartheta}$$

$$\Gamma^1_{33} = -\frac{r \sin^2 \vartheta (r^2 + a^2)}{r^2 + a^2 \cos^2 \vartheta}$$

$$\Gamma^2_{33} = -\frac{\sin \vartheta \cos \vartheta (r^2 + a^2)}{r^2 + a^2 \cos^2 \vartheta}$$

$$\Gamma^3_{31} = \Gamma^3_{13} = \frac{r}{r^2 + a^2}$$

$$\Gamma^3_{32} = \Gamma^3_{23} = \frac{\cos \vartheta}{\sin \vartheta}$$

$$\Gamma^k_{ij} = 0, \text{ otherwise.}$$

The components of the Ricci tensor of the limit A.2.4 Now we calculate the components of the Ricci tensor of the limit. The following holds:

$$\text{Ric}_{00} = \partial_j \Gamma^j_{00} - \partial_0 \Gamma^j_{j0} + \Gamma^r_{jr} \Gamma^j_{00} - \Gamma^j_{0r} \Gamma^r_{j0} = \partial_1 \Gamma^1_{00} + \partial_2 \Gamma^2_{00} + \Gamma^1_{00}(\Gamma^1_{11} + \Gamma^2_{21} + \Gamma^3_{31})$$

$$+ \Gamma^2_{00}(\Gamma^1_{12} + \Gamma^2_{22} + \Gamma^3_{32}) = \frac{1}{\rho^4}(-2r^5 + 4r^3 a^2 \cos^2 \vartheta - 4r^3 a^2 \sin^2 \vartheta + 2ra^4 \sin^2 \vartheta \cos^2 \vartheta$$

$$+ 6ra^4 \cos^2 \vartheta - 2r^3 a^2 \cos^2 \vartheta + 2r^3 a^2 \sin^2 \vartheta - 2ra^4 \cos^4 \vartheta - 10ra^4 \sin^2 \vartheta \cos^2 \vartheta)$$

$$+ \frac{(r^2 + a^2)(r^2 - a^2 \cos^2 \vartheta)}{\rho^3} \left(\frac{ra^2 \sin^2 \vartheta}{\rho(r^2 + a^2)} + \frac{r}{\rho} + \frac{r}{r^2 + a^2} \right)$$

$$- \frac{2ra^2 \sin \vartheta \cos \vartheta}{\rho^3} \left(-\frac{2a^2 \sin \vartheta \cos \vartheta}{\rho} + \frac{\cos \vartheta}{\sin \vartheta} \right)$$

$$= \frac{1}{\rho^4}(-2r^5 + 2r^3 a^2 \cos^2 \vartheta - 2r^3 a^2 \sin^2 \vartheta - 8ra^4 \sin^2 \vartheta \cos^2 \vartheta + 6ra^4 \cos^4 \vartheta$$

$$- 2ra^4 \cos^4 \vartheta + r^3 a^2 \sin^2 \vartheta + (r^3 + ra^2)(r^2 - a^2 \cos^2 \vartheta) + r(r^4 - a^4 \cos^4 \vartheta)$$

$$+ 4ra^4 \sin^2 \vartheta \cos^2 \vartheta - 2ra^2 \cos^2 \vartheta (r^2 + a^2 \cos^2 \vartheta)$$

$$= \frac{1}{\rho^4}(-2r^5 + 2r^3a^2\cos^2\vartheta - r^3a^2\sin^2\vartheta - 5ra^4\sin^2\vartheta\cos^2\vartheta + 6ra^4\cos^2\vartheta$$
$$- 2ra^4\cos^4\vartheta + r^5 + r^3a^2 - r^3a^2\cos^2\vartheta - ra^4\cos^2\vartheta + r^5 - ra^4\cos^4\vartheta$$
$$- 2r^3a^2\cos^2\vartheta - 2ra^4\cos^4\vartheta)$$
$$= \frac{1}{\rho^4}(-5ra^4\cos^4\vartheta - 5ra^4\sin^2\vartheta\cos^2\vartheta + 5ra^4\cos^2\vartheta)$$
$$= \frac{-5ra^4\cos^2\vartheta}{\rho^4}(\sin^2\vartheta + \cos^2\vartheta - 1) = 0$$

$\mathrm{Ric}_{01} = \mathrm{Ric}_{10} = \partial_j\Gamma^j_{01} - \partial_0\Gamma^j_{j1} + \Gamma^r_{jr}\Gamma^r_{01} - \Gamma^j_{0r}\Gamma^r_{j1} = 0$

$\mathrm{Ric}_{02} = \mathrm{Ric}_{20} = \partial_j\Gamma^j_{02} - \partial_0\Gamma^j_{j2} + \Gamma^r_{jr}\Gamma^r_{02} - \Gamma^j_{0r}\Gamma^r_{j2} = 0$

$\mathrm{Ric}_{03} = \mathrm{Ric}_{30} = \partial_j\Gamma^j_{03} - \partial_0\Gamma^j_{j3} + \Gamma^r_{jr}\Gamma^r_{03} - \Gamma^j_{0r}\Gamma^r_{j3} = 0$

$\mathrm{Ric}_{11} = \partial_j\Gamma^j_{11} - \partial_1\Gamma^j_{j1} + \Gamma^r_{jr}\Gamma^r_{11} - \Gamma^j_{1r}\Gamma^r_{j1}$
$$= \partial_1\Gamma^1_{11} + \partial_2\Gamma^2_{11} - \partial_1\Gamma^1_{11} - \partial_1\Gamma^2_{21} - \partial_1\Gamma^3_{31} + \Gamma^1_{11}(\Gamma^1_{11} + \Gamma^2_{21} + \Gamma^3_{31})$$
$$+ \Gamma^2_{11}(\Gamma^1_{12} + \Gamma^2_{22} + \Gamma^3_{32}) - \Gamma^1_{11}\Gamma^1_{11} - \Gamma^2_{11}\Gamma^1_{21} - \Gamma^1_{12}\Gamma^2_{11} - \Gamma^2_{12}\Gamma^2_{21} - \Gamma^3_{13}\Gamma^3_{31}$$
$$= \frac{a^2\cos^2\vartheta(r^2+a^2) - a^2r^2\sin^2\vartheta}{\rho^2(r^2+a^2)} - \frac{-r^2+a^2\cos^2\vartheta}{\rho^2} - \frac{-r^2+a^2}{(r^2+a^2)^2}$$
$$+ \frac{ra\sin^2\vartheta}{\rho(r^2+a^2)}\left(\frac{r}{\rho} + \frac{r}{(r^2+a^2)}\right) + \frac{a^2\sin\vartheta\cos\vartheta}{\rho(r^2+a^2)} \cdot \frac{\cos\vartheta}{\sin\vartheta} - \frac{r^2}{\rho^2} - \frac{r^2}{(r^2+a^2)^2}$$
$$= \frac{1}{\rho^2(r^2+a^2)^2}[-a^2r^4\sin^2\vartheta - a^2(r^4 + 2r^2a^2\cos^2\vartheta + a^4\cos^4\vartheta)$$
$$+ 2r^4a^2\sin^2\vartheta + r^2a^4\sin^2\vartheta\cos^2\vartheta + (r^2a^2\cos^2\vartheta + a^4\cos^4\vartheta)(r^2+a^2)]$$
$$= \frac{1}{\rho^2(r^2+a^2)^2}[-a^2r^4 - 2r^2a^4\cos^2\vartheta - a^6\cos^4\vartheta + r^4a^2\sin^2\vartheta$$
$$+ r^2a^4\sin^2\vartheta\cos^2\vartheta + r^4a^2\cos^2\vartheta + r^2a^4\cos^2\vartheta + a^4r^2\cos^4\vartheta + a^6\cos^4\vartheta]$$
$$= \frac{1}{\rho^2(r^2+a^2)^2}[-a^2r^4 - 2r^2a^4\cos^2\vartheta + r^4a^2 - r^4a^2\cos^2\vartheta + r^2a^4\cos^2\vartheta$$
$$- r^2a^4\cos^4\vartheta + r^4a^2\cos^2\vartheta + r^2a^4\cos^2\vartheta + a^4r^2\cos^4\vartheta] = 0$$

$\mathrm{Ric}_{21} = \mathrm{Ric}_{12} = \partial_j\Gamma^j_{21} - \partial_2\Gamma^j_{j1} + \Gamma^r_{jr}\Gamma^r_{21} - \Gamma^j_{2r}\Gamma^r_{j1}$
$$= \partial_1\Gamma^1_{21} + \partial_2\Gamma^2_{21} - \partial_2\Gamma^1_{11} - \partial_2\Gamma^2_{21} - \partial_2\Gamma^3_{31} + \Gamma^2_{21}(\Gamma^1_{12} + \Gamma^2_{22} + \Gamma^3_{32})$$
$$+ \Gamma^1_{21}(\Gamma^1_{11} + \Gamma^2_{21} + \Gamma^3_{31}) - \Gamma^1_{21}\Gamma^1_{11} - \Gamma^2_{21}\Gamma^1_{21} - \Gamma^1_{22}\Gamma^2_{11} - \Gamma^2_{22}\Gamma^2_{21} - \Gamma^3_{23}\Gamma^3_{31}$$
$$= \frac{2ra^2\sin\vartheta\cos\vartheta}{\rho^2} - \frac{2ra^2\sin\vartheta\cos\vartheta}{\rho^2} + \frac{r}{\rho} \cdot \frac{\cos\vartheta}{\sin\vartheta} - \frac{a^2\sin\vartheta\cos\vartheta}{\rho^2}\left(\frac{r}{\rho} + \frac{r}{r^2+a^2}\right)$$
$$+ \frac{r(a^2+r^2)}{\rho} \cdot \frac{a^2\sin\vartheta\cos\vartheta}{\rho(r^2+a^2)} - \frac{r}{r^2+a^2} \cdot \frac{\cos\vartheta}{\sin\vartheta}$$
$$= \frac{\cos\vartheta}{\sin\vartheta}\left(\frac{r}{\rho} - \frac{r}{r^2+a^2}\right) - \frac{2r^3a^2\sin\vartheta\cos\vartheta + ra^4\sin\vartheta\cos\vartheta + ra^4\sin\vartheta\cos^3\vartheta}{\rho^2(r^2+a^2)}$$
$$+ \frac{ra^4\sin\vartheta\cos\vartheta + r^3a^2\sin\vartheta\cos\vartheta}{\rho^2(r^2+a^2)}$$
$$= \frac{r^3a^2\sin\vartheta\cos\vartheta + ra^4\sin\vartheta\cos^3\vartheta}{\rho^2(r^2+a^2)} - \frac{r^3a^2\sin\vartheta\cos\vartheta + ra^4\sin\vartheta\cos^3\vartheta}{\rho^2(r^2+a^2)} = 0$$

$\mathrm{Ric}_{31} = \mathrm{Ric}_{13} = \partial_j\Gamma^j_{31} - \partial_3\Gamma^j_{j1} + \Gamma^r_{jr}\Gamma^r_{31} - \Gamma^j_{3r}\Gamma^r_{j1} = 0$

A.2. The Kerr spacetime

$$\text{Ric}_{22} = \partial_j \Gamma^j_{22} - \partial_2 \Gamma^j_{j2} + \Gamma^r_{jr} \Gamma^j_{22} - \Gamma^j_{2r} \Gamma^r_{j2}$$

$$= \partial_1 \Gamma^1_{22} + \partial_2 \Gamma^2_{22} - \partial_2 \Gamma^1_{12} - \partial_2 \Gamma^2_{22} - \partial_2 \Gamma^3_{32} + \Gamma^1_{22}(\Gamma^1_{11} + \Gamma^2_{21} + \Gamma^3_{31})$$

$$+ \Gamma^2_{22}(\Gamma^1_{12} + \Gamma^2_{22} + \Gamma^3_{32}) - \Gamma^1_{21}\Gamma^1_{12} - \Gamma^2_{21}\Gamma^1_{22} - \Gamma^1_{22}\Gamma^2_{12} - \Gamma^2_{22}\Gamma^2_{22} - \Gamma^3_{23}\Gamma^3_{32}$$

$$= \frac{-r^4 + r^2 a^2 - 3 r^2 a^2 \cos^2 \vartheta - a^2 \cos^2 \vartheta}{\rho^2} - \frac{r^2 a^2 \sin^2 \vartheta - r^2 a^2 \cos^2 \vartheta - a^4 \cos^2 \vartheta}{\rho^2}$$

$$+ \frac{1}{\sin^2 \vartheta} - \frac{r^2(r^2 + a^2)}{\rho} \left(\frac{r a^2 \sin^2 \vartheta}{\rho(r^2 + a^2)} + \frac{r}{r^2 + a^2} \right) - \frac{a^2 \cos \vartheta \sin \vartheta}{\rho} \cdot \frac{\cos \vartheta}{\sin \vartheta}$$

$$+ \frac{r^2(r^2 + a^2)}{\rho^2} - \frac{\cos^2 \vartheta}{\sin^2 \vartheta}$$

$$= \frac{1}{\rho^2}[-r^4 + r^2 a^2 - 2 r^2 a^2 \cos^2 \vartheta - r^2 a^2 \sin^2 \vartheta - r^2 a^2 \sin^2 \vartheta - r^4 - r^2 a^2 \cos^2 \vartheta$$

$$- a^2 \cos^2 \vartheta (r^2 + a^2 \cos^2 \vartheta) + (r^2 + a^2 \cos^2 \vartheta)^2 + a^2 r^2 + r^4]$$

$$= \frac{1}{\rho^2}[2 r^2 a^2 - 3 r^2 a^2 \cos^2 \vartheta - 2 r^2 a^2 \sin^2 \vartheta - a^2 r^2 \cos^2 \vartheta - a^4 \cos^4 \vartheta - r^4 + r^4$$

$$+ 2 r^2 a^2 \cos^2 \vartheta + a^4 \cos^4 \vartheta] = \frac{1}{\rho^2}[2 r^2 a^2 - 2 r^2 a^2 (\cos^2 \vartheta + \sin^2 \vartheta)] = 0$$

$$\text{Ric}_{23} = \text{Ric}_{32} = \partial_j \Gamma^j_{23} - \partial_2 \Gamma^j_{j3} + \Gamma^r_{jr} \Gamma^j_{23} - \Gamma^j_{2r} \Gamma^r_{j3} = 0$$

$$\text{Ric}_{33} = \partial_j \Gamma^j_{33} - \partial_3 \Gamma^j_{j3} + \Gamma^r_{jr} \Gamma^j_{33} - \Gamma^j_{3r} \Gamma^r_{j3} = \partial_1 \Gamma^1_{33} + \partial_2 \Gamma^2_{33} + \Gamma^1_{33}(\Gamma^1_{11} + \Gamma^2_{21} + \Gamma^3_{31})$$

$$+ \Gamma^2_{33}(\Gamma^1_{12} + \Gamma^2_{22} + \Gamma^3_{32}) - \Gamma^1_{33}\Gamma^3_{13} - \Gamma^3_{33}\Gamma^3_{23} - \Gamma^3_{31}\Gamma^1_{33} - \Gamma^3_{32}\Gamma^2_{33}$$

$$= \frac{-r^4 \sin^2 \vartheta - a^4 \sin^2 \vartheta - 3 r^2 a^2 \sin^2 \vartheta \cos^2 \vartheta + r^2 a^2 \sin^2 \vartheta}{\rho^2}$$

$$+ \frac{-r^4 \cos^2 - r^2 a^2 \cos^4 \vartheta - r^2 a^2 \cos^2 \vartheta - a^4 \cos^4 \vartheta}{\rho^2} + \frac{r^4 \sin^2 \vartheta + r^2 a^2 \sin^2 \vartheta}{\rho^2}$$

$$- \frac{r \sin^2 \vartheta (r^2 + a^2)}{\rho} \left(\frac{r a^2 \sin^2 \vartheta}{\rho(r^2 + a^2)} + \frac{r^3 + r a^2}{\rho(r^2 + a^2)} \right)$$

$$- \frac{\sin \vartheta \cos \vartheta (r^2 + a^2)}{\rho} \left(-\frac{2 a^2 \cos \vartheta \sin \vartheta}{\rho} \right)$$

$$+ \frac{r \sin^2 \vartheta (r^2 + a^2)}{\rho} \cdot \frac{r}{r^2 + a^2} + \frac{\sin \vartheta \cos \vartheta (r^2 + a^2)}{\rho} \cdot \frac{\cos \vartheta}{\sin \vartheta}$$

$$= \frac{1}{\rho^2}[-r^4 \sin^2 \vartheta - a^4 \sin^2 \vartheta \cos^2 \vartheta - 3 r^2 a^2 \sin^2 \vartheta \cos^2 \vartheta + r^2 a^2 \sin^2 \vartheta$$

$$- r^2 a^2 \cos^4 \vartheta - r^2 a^2 \cos^2 \vartheta - a^4 \cos^4 \vartheta + r^4 \sin^2 \vartheta - r^4 \cos^2 \vartheta + r^2 a^2 \sin^2 \vartheta$$

$$- r^2 a^2 \sin^2 \vartheta \cos^2 \vartheta - a^4 \sin^2 \vartheta \cos^2 \vartheta - r^2 a^2 \sin^4 \vartheta - r^4 \sin^2 \vartheta - r^2 a^2 \sin^2 \vartheta$$

$$+ 2 r^2 a^2 \sin^2 \vartheta \cos^2 \vartheta + 2 a^4 \sin^2 \vartheta \cos^2 \vartheta + r^4 \sin^2 \vartheta + r^2 a^2 \sin^2 \vartheta \cos^2 \vartheta$$

$$+ r^4 \cos^2 \vartheta + r^2 a^2 \cos^4 \vartheta + a^2 r^2 \cos^2 \vartheta + a^4 \cos^4 \vartheta]$$

$$= \frac{1}{\rho^2}[-r^2 a^2 \sin^2 \vartheta \cos^2 \vartheta + r^2 a^2 \sin^2 \vartheta - r^2 a^2 \sin^4 \vartheta]$$

$$= \frac{1}{\rho^2}[-r^2 a^2 \sin^2 \vartheta \cos^2 \vartheta + r^2 a^2 \sin^2 \vartheta - r^2 a^2 \sin^2 \vartheta + r^2 a^2 \sin^2 \vartheta \cos^2 \vartheta] = 0.$$

The components of the curvature tensor A.2.5 Now we calculate the components of the curvature tensor, $R^l_{ijk} = \partial_j \Gamma^l_{ki} - \partial_k \Gamma^l_{ji} + \Gamma^l_{jr} \Gamma^r_{ki} - \Gamma^l_{kr} \Gamma^r_{ij}$. As the connection is symmetric,

$R^l_{ijk} = -R^l_{ikj}$ also holds. We only care for the components which do not vanish immediately. First, we have to calculate the derivations of the Christoffel symbols - of course, only for $i = 1$ (which means for r) and for $i = 2$ (which means for ϑ). As we are only interested in the asymptotical behaviour of the components in r, we just examine this behaviour. We then get:

$$\partial_1 \Gamma^1_{00} = \partial_r \left(\frac{r^4 - 2\lambda r^3 + a^2 r^2 \sin^2 \vartheta + 2\lambda r a^2 \cos^2 \vartheta - a^4 \cos^2 \vartheta}{(r^2 + a^2 \cos^2 \vartheta)^3} \right)$$
$$= \frac{4r^3 - 6\lambda r^2 + 2ra^2 \sin^2 \vartheta + 2\lambda a^2 \cos^2 \vartheta}{\rho^3}$$
$$- \frac{(r^4 - 2\lambda r^3 + r^2 a^2 \sin^2 \vartheta + 2\lambda r a^2 \cos^2 \vartheta - a^4 \cos^2 \vartheta) \cdot 3 \cdot 2r}{\rho^4} = \mathcal{O}\left(\frac{1}{r^3}\right)$$

$$\partial_2 \Gamma^1_{00} = \widetilde{\Delta} \cdot \frac{\rho \cdot 2a^2 \sin \vartheta \cos \vartheta + (r^2 - a^2 \cos^2 \vartheta) \cdot 3 \cdot 2a^2 \cos \vartheta \sin \vartheta}{\rho^4} = \mathcal{O}\left(\frac{1}{r^4}\right)$$

$$\partial_1 \Gamma^2_{00} = \frac{-2a^2 \cos^2 \vartheta \cdot \rho + 2ra^2 \sin \vartheta \cos \vartheta \cdot 3 \cdot 2r}{\rho^4} = \mathcal{O}\left(\frac{1}{r^6}\right)$$

$$\partial_2 \Gamma^2_{00} = \frac{\rho \cdot (-2ra^2 \cos^2 \vartheta + 2ra^2 \sin^2 \vartheta) + 2ra^2 \sin \vartheta \cos \vartheta \cdot 3 \cdot (-2a^2 \cos \vartheta \sin \vartheta)}{\rho^4} = \mathcal{O}\left(\frac{1}{r^5}\right)$$

$$\partial_1 \Gamma^0_{01} = \lambda \cdot \frac{(4r^3 + 2ra^2 \sin^2 \vartheta)\rho\widetilde{\Delta} - (r^4 + r^2 a^2 \sin^2 \vartheta - a^4 \cos^2 \vartheta)(4r\widetilde{\Delta} + \rho(2r - 2\lambda))}{\rho^3 \widetilde{\Delta}^2} = \mathcal{O}\left(\frac{1}{r^3}\right)$$

$$\partial_2 \Gamma^0_{01} = \frac{\lambda(r^2 + a^2)}{\widetilde{\Delta}} \cdot \frac{2a^2 \sin \vartheta \cos \vartheta \cdot \rho - (r^2 - a^2 \cos^2 \vartheta) \cdot 2 \cdot (-2a^2 \sin \vartheta \cos \vartheta)}{\rho^3} = \mathcal{O}\left(\frac{1}{r^4}\right)$$

$$\partial_1 \Gamma^0_{02} = \frac{-4\lambda a^2 \sin \vartheta \cos \vartheta \cdot \rho + 4\lambda r a^2 \sin \vartheta \cos \vartheta \cdot 2 \cdot 2r}{\rho^3} = \mathcal{O}\left(\frac{1}{r^4}\right)$$

$$\partial_2 \Gamma^0_{02} = \frac{(-4\lambda r a^2 \cos^2 \vartheta + 4\lambda r a^2 \sin^2 \vartheta)\rho + 4\lambda r a^2 \sin \vartheta \cos \vartheta \cdot 2 \cdot (-2a^2 \sin \vartheta \cos \vartheta)}{\rho^3} = \mathcal{O}\left(\frac{1}{r^3}\right)$$

$$\partial_1 \Gamma^3_{01} = \sqrt{\lambda} a \cdot \frac{2r\widetilde{\Delta}\rho - (r^2 - a^2 \cos^2 \vartheta)((2\lambda - 2r)\rho + 2 \cdot 2r\widetilde{\Delta})}{\widetilde{\Delta}^2 \rho^3} = \mathcal{O}\left(\frac{1}{r^5}\right)$$

$$\partial_2 \Gamma^3_{01} = \frac{2\sqrt{\lambda} a^3 \sin \vartheta \cos \vartheta (3r^2 - a^2 \cos^2 \vartheta)}{\widetilde{\Delta}\rho^3} = \mathcal{O}\left(\frac{1}{r^6}\right)$$

$$\partial_1 \Gamma^3_{02} = \frac{2\sqrt{\lambda} a \cos \vartheta (3r^2 - a^2 \cos^2 \vartheta)}{\sin \vartheta \rho^3} = \mathcal{O}\left(\frac{1}{r^4}\right)$$

$$\partial_2 \Gamma^3_{02} = \frac{-2\sqrt{\lambda} r a(-\rho + 4a^2 \sin^2 \vartheta \cos^2 \vartheta)}{\sin^2 \vartheta \rho^3} = \mathcal{O}\left(\frac{1}{r^3}\right)$$

$$\partial_1 \Gamma^0_{13} = \left(\frac{(4r^3 + 2ra^2 \sin^2 \vartheta)\rho\widetilde{\Delta} - (r^4 + r^2 a^2 \sin^2 \vartheta - a^4 \cos^2 \vartheta)(4r\widetilde{\Delta} + \rho(2r - 2\lambda))}{\rho^3 \widetilde{\Delta}^2} \right.$$
$$\left. + \frac{2\widetilde{\Delta}\rho - 2r^2(2r\widetilde{\Delta} + (2r - 2\lambda)\rho)}{\rho^2 \widetilde{\Delta}^2} \right) \cdot (-\lambda^{\frac{3}{2}} a \sin^2 \vartheta) = \mathcal{O}\left(\frac{1}{r^3}\right)$$

$$\partial_2 \Gamma^0_{13} = \frac{-\lambda^{\frac{3}{2}} a}{\widetilde{\Delta}} \left(\frac{(r^2 + a^2)(2 \sin \vartheta \cos \vartheta (r^2 - a^2 \cos^2 \vartheta) + \sin^2 \vartheta (2a^2 \sin \vartheta \cos \vartheta))}{\rho^2} \right.$$
$$\left. - \frac{\sin^2 \vartheta (r^2 - a^2 \cos^2 \vartheta) \cdot 2 \cdot (-2a^2 \cos \vartheta \sin \vartheta)}{\rho^3} - \frac{2r^2(-2a^2 \sin \vartheta \cos \vartheta)}{\rho^2} \right) = \mathcal{O}\left(\frac{1}{r^2}\right)$$

$$\partial_1 \Gamma^0_{23} = \frac{2\lambda^{\frac{3}{2}} a^2 \sin^3 \vartheta \cos \vartheta (-3r^2 + a^2 \cos^2 \vartheta)}{\rho^3} = \mathcal{O}\left(\frac{1}{r^4}\right)$$

$$\partial_2 \Gamma^0_{23} = \frac{2\lambda^{\frac{3}{2}} a^2 r ((3\sin^2 \vartheta \cos^2 \vartheta - \sin^4 \vartheta)\rho - \sin^3 \vartheta \cos \vartheta \cdot 2(-2a^2 \sin \vartheta \cos \vartheta))}{\rho^3} = \mathcal{O}\left(\frac{1}{r^3}\right)$$

$$\partial_1 \Gamma^1_{30} = -\sqrt{\lambda} a \sin^2 \vartheta \frac{((2r - 2\lambda)(r^2 - a^2 \cos^2 \vartheta) + 2r \cdot \widetilde{\Delta})\rho - (r^2 - a^2 \cos^2 \vartheta)\widetilde{\Delta} \cdot 3 \cdot 2r}{\rho^4}$$
$$= \mathcal{O}\left(\frac{1}{r^3}\right)$$

$$\partial_2 \Gamma^1_{30} = -\sqrt{\lambda} a \cdot \widetilde{\Delta} \left(\frac{2\sin \vartheta \cos \vartheta (r^2 - a^2 \cos^2 \vartheta) + \sin^2 \vartheta \cdot 2a^2 \sin \vartheta \cos \vartheta}{\rho^3} \right.$$
$$\left. - \frac{\sin^2 \vartheta (r^2 - a^2 \cos^2 \vartheta) \cdot 3(-2a^2 \sin \vartheta \cos \vartheta)}{\rho^4} \right) = \mathcal{O}\left(\frac{1}{r^2}\right)$$

$$\partial_1 \Gamma^2_{30} = 2\sqrt{\lambda} a \sin \vartheta \cos \vartheta \frac{(3r^2 + a^2)\rho - (r^3 + ra^2) \cdot 3 \cdot 2r}{\rho^4} = \mathcal{O}\left(\frac{1}{r^4}\right)$$

$$\partial_2 \Gamma^2_{30} = 2\sqrt{\lambda} ra(r^2 + a^2) \frac{(\cos^2 \vartheta - \sin^2 \vartheta)\rho - \sin \vartheta \cos \vartheta \cdot 3(-2a^2 \sin \vartheta \cos \vartheta)}{\rho^4} = \mathcal{O}\left(\frac{1}{r^3}\right)$$

$$\partial_1 \Gamma^1_{11} = \frac{(a^2 \sin^2 \vartheta - 2\lambda r)\widetilde{\Delta}\rho - (ra^2 \sin^2 \vartheta - \lambda(r^2 - a^2 \cos^2 \vartheta))(2r \cdot \widetilde{\Delta} + (2r - 2\lambda)\rho)}{\widetilde{\Delta}^2 \rho^2} = \mathcal{O}\left(\frac{1}{r^3}\right)$$

$$\partial_2 \Gamma^1_{11} = \frac{2a^2 \sin \vartheta \cos \vartheta (r - \lambda)\rho - (ra^2 \sin^2 \vartheta - \lambda(r^2 - a^2 \cos^2 \vartheta))(-2a^2 \sin \vartheta \cos \vartheta)}{\widetilde{\Delta}\rho^2} = \mathcal{O}\left(\frac{1}{r^3}\right)$$

$$\partial_1 \Gamma^2_{11} = \frac{-a^2 \sin \vartheta \cos \vartheta (2r\widetilde{\Delta} + (2r - 2\lambda)\rho)}{\rho^2 \widetilde{\Delta}^2} = \mathcal{O}\left(\frac{1}{r^5}\right)$$

$$\partial_2 \Gamma^2_{11} = \frac{(a^2 \cos^2 \vartheta - a^2 \sin^2 \vartheta)\rho - a^2 \sin \vartheta \cos \vartheta (-2a^2 \sin \vartheta \cos \vartheta)}{\widetilde{\Delta}\rho^2} = \mathcal{O}\left(\frac{1}{r^4}\right)$$

$$\partial_1 \Gamma^1_{21} = \frac{2ra^2 \sin \vartheta \cos \vartheta}{\rho^2} = \mathcal{O}\left(\frac{1}{r^3}\right)$$

$$\partial_2 \Gamma^1_{21} = \frac{-(-a^2 \sin \vartheta + a^2 \cos \vartheta)\rho + a^2 \sin \vartheta \cos \vartheta (-2a^2 \sin \vartheta \cos \vartheta)}{\rho^2} = \mathcal{O}\left(\frac{1}{r^2}\right)$$

$$\partial_1 \Gamma^2_{22} = \partial_1 \Gamma^1_{21} = \mathcal{O}\left(\frac{1}{r^3}\right)$$

$$\partial_2 \Gamma^2_{22} = \partial_2 \Gamma^1_{21} = \mathcal{O}\left(\frac{1}{r^2}\right)$$

$$\partial_1 \Gamma^1_{22} = \frac{(-3r^2 + 4\lambda r - a^2)\rho - (-r^3 + 2\lambda r^2 - ra^2) \cdot 2r}{\rho^2} = \text{const.}$$

$$\partial_2 \Gamma^1_{22} = \frac{-2ra^2 \sin \vartheta \cos \vartheta \cdot \widetilde{\Delta}}{\rho^2} = \mathcal{O}\left(\frac{1}{r}\right)$$

$$\partial_1 \Gamma^2_{21} = \frac{a^2 \cos^2 \vartheta - r^2}{\rho^2} = \mathcal{O}\left(\frac{1}{r^2}\right)$$

$$\partial_2 \Gamma^2_{21} = \frac{2ra^2 \sin \vartheta \cos \vartheta}{\rho^2} = \mathcal{O}\left(\frac{1}{r^3}\right)$$

$$\partial_1 \Gamma^1_{33} = -\sin^2 \vartheta \left(\frac{((2r - 2\lambda)r + \widetilde{\Delta})\rho - r \cdot 2r \cdot \widetilde{\Delta}}{\rho^2} - \lambda a^2 \sin^2 \vartheta \cdot \right.$$
$$\left. \frac{\rho((2r - 2\lambda)(r^2 - a^2 \cos^2 \vartheta) + 2r \cdot \widetilde{\Delta}) - 2 \cdot 2r(r^2 - a^2 \cos^2 \vartheta)\widetilde{\Delta}}{\rho^3} \right) = \text{const.}$$

A.2. The Kerr spacetime

$$\partial_2 \Gamma^1_{33} = -\widetilde{\Delta}\left(\frac{2r\sin\vartheta\cos\vartheta\cdot\rho - r\sin^2\vartheta(-2a^2\sin\vartheta\cos\vartheta)}{\rho^2}\right.$$
$$-\frac{\lambda a^2(4\sin^3\vartheta\cos\vartheta(r^2-a^2\cos^2\vartheta)+a^2\sin^4\vartheta(2a^2\sin\vartheta\cos\vartheta))}{\rho^3}$$
$$\left.+\frac{\lambda a^2\sin^4\vartheta(r^2-a^2\cos^2\vartheta)\cdot 3(-2a^2\sin\vartheta\cos\vartheta)}{\rho^4}\right) = \mathcal{O}(r)$$

$$\partial_1 \Gamma^2_{33} = -\sin\vartheta\cos\vartheta\left(\frac{(2r-2\lambda)\rho - 2r\widetilde{\Delta}}{\rho^2}\right.$$
$$\left.+\frac{(10\lambda r^4+12\lambda r^2 a^2+2\lambda a^4)\rho - (2\lambda r^5+4\lambda r^3 a^2+2\lambda r a^4)\cdot 3\cdot 2r}{\rho^4}\right) = \mathcal{O}\left(\frac{1}{r^2}\right)$$

$$\partial_2 \Gamma^2_{33} = -\widetilde{\Delta}\cdot\frac{(\cos^2\vartheta-\sin^2\vartheta)\rho - \sin\vartheta\cos\vartheta(-2a^2\sin\vartheta\cos\vartheta)}{\rho^2}$$
$$-2\lambda r(r^2+a^2)^2\cdot\frac{-2(-2a^2\sin\vartheta\cos\vartheta)}{\rho^4} = \text{const.}$$

$$\partial_1 \Gamma^3_{31} = -2\lambda\cdot\frac{2r\cdot\widetilde{\Delta}\rho - r^2((2r-2\lambda)\rho + 2r\cdot\widetilde{\Delta})}{\widetilde{\Delta}^2\rho^2} + \lambda a^2\sin^2\vartheta\cdot$$
$$\frac{2r\cdot\widetilde{\Delta}\rho - (r^2-a^2\cos^2\vartheta)((2r-2\lambda)\rho+2\cdot 2r\cdot\widetilde{\Delta})}{\widetilde{\Delta}^2\rho^3} + \frac{\widetilde{\Delta}-r(2r-2\lambda)}{\widetilde{\Delta}^2} = \mathcal{O}\left(\frac{1}{r^2}\right)$$

$$\partial_2 \Gamma^3_{31} = \frac{1}{\widetilde{\Delta}}\left(-2\lambda r^2\cdot\frac{2a^2\sin\vartheta\cos\vartheta}{\rho^2}\right.$$
$$+\lambda a^2\cdot\frac{2\sin\vartheta\cos\vartheta(r^2-a^2\cos^2\vartheta)+\sin^2\vartheta(2a^2\sin\vartheta\cos\vartheta)}{\rho^2}$$
$$\left.-\frac{\sin^2\vartheta(r^2-a^2\cos^2\vartheta)\cdot 2(-2a^2\sin\vartheta\cos\vartheta)}{\rho^3}\right) = \mathcal{O}\left(\frac{1}{r^4}\right)$$

$$\partial_1 \Gamma^3_{32} = \frac{\cos\vartheta}{\sin\vartheta}\cdot 2\lambda a^2\sin^2\vartheta\cdot\frac{\rho - r\cdot 2\cdot 2r}{\rho^3} = \mathcal{O}\left(\frac{1}{r^4}\right)$$

$$\partial_2 \Gamma^3_{32} = 2\lambda r a^2\cdot\frac{(\cos^2\vartheta-\sin^2\vartheta)\rho-\sin\vartheta\cos\vartheta\cdot 2(-2a^2\sin\vartheta\cos\vartheta)}{\rho^3} - \frac{1}{\sin^2\vartheta} = \text{const.}$$

Now we are able to calculate the asymptotical behaviour in r of the curvature tensor of the extended Kerr spacetime. We only consider those components which do not vanish immediately. (Remember: $R^l_{ijk} = \partial_j \Gamma^l_{ki} - \partial_k \Gamma^l_{ji} + \Gamma^l_{jr}\Gamma^r_{ki} - \Gamma^l_{kr}\Gamma^r_{ij}$, and as the connection is symmetric, $R^l_{ijk} = -R^l_{ikj}$ also holds.) We then get:

$$R^0_{030} = -R^0_{003} = \Gamma^0_{31}\Gamma^1_{00} + \Gamma^0_{32}\Gamma^2_{00} - \Gamma^0_{01}\Gamma^1_{03} - \Gamma^0_{02}\Gamma^2_{03}$$
$$= \mathcal{O}\left(\frac{1}{r^4}\right) + \mathcal{O}\left(\frac{1}{r^8}\right) + \mathcal{O}\left(\frac{1}{r^4}\right) + \mathcal{O}\left(\frac{1}{r^6}\right) = \mathcal{O}\left(\frac{1}{r^4}\right)$$

$$R^0_{110} = -R^0_{101} = \partial_1\Gamma^0_{01} + \Gamma^0_{10}\Gamma^0_{01} + \Gamma^0_{13}\Gamma^3_{01} - \Gamma^0_{01}\Gamma^1_{11} - \Gamma^0_{02}\Gamma^2_{11}$$
$$= \mathcal{O}\left(\frac{1}{r^3}\right) + \mathcal{O}\left(\frac{1}{r^4}\right) + \mathcal{O}\left(\frac{1}{r^6}\right) + \mathcal{O}\left(\frac{1}{r^4}\right) + \mathcal{O}\left(\frac{1}{r^7}\right) = \mathcal{O}\left(\frac{1}{r^3}\right)$$

$$R^0_{220} = -R^0_{202} = \partial_2\Gamma^0_{02} + \Gamma^0_{20}\Gamma^0_{02} + \Gamma^0_{23}\Gamma^3_{02} - \Gamma^0_{01}\Gamma^1_{22} - \Gamma^0_{02}\Gamma^2_{22}$$
$$= \mathcal{O}\left(\frac{1}{r^3}\right) + \mathcal{O}\left(\frac{1}{r^6}\right) + \mathcal{O}\left(\frac{1}{r^6}\right) + \mathcal{O}\left(\frac{1}{r}\right) + \mathcal{O}\left(\frac{1}{r^5}\right) = \mathcal{O}\left(\frac{1}{r}\right)$$

A.2. The Kerr spacetime

$$R^0_{330} = -R^0_{303} = \Gamma^0_{31}\Gamma^1_{03} + \Gamma^0_{32}\Gamma^2_{03} - \Gamma^0_{01}\Gamma^1_{33} - \Gamma^0_{02}\Gamma^2_{33}$$
$$= \mathcal{O}\left(\frac{1}{r^4}\right) + \mathcal{O}\left(\frac{1}{r^6}\right) + \mathcal{O}\left(\frac{1}{r}\right) + \mathcal{O}\left(\frac{1}{r^3}\right) = \mathcal{O}\left(\frac{1}{r}\right)$$

$$R^0_{120} = -R^0_{102} = \partial_2\Gamma^0_{01} + \Gamma^0_{20}\Gamma^0_{01} + \Gamma^0_{23}\Gamma^3_{01} - \Gamma^0_{01}\Gamma^1_{12} - \Gamma^0_{02}\Gamma^2_{12}$$
$$= \mathcal{O}\left(\frac{1}{r^4}\right) + \mathcal{O}\left(\frac{1}{r^5}\right) + \mathcal{O}\left(\frac{1}{r^7}\right) + \mathcal{O}\left(\frac{1}{r^4}\right) + \mathcal{O}\left(\frac{1}{r^4}\right) = \mathcal{O}\left(\frac{1}{r^4}\right)$$

$$R^0_{210} = -R^0_{201} = \partial_1\Gamma^0_{02} + \Gamma^0_{10}\Gamma^0_{02} + \Gamma^0_{13}\Gamma^3_{02} - \Gamma^0_{01}\Gamma^1_{21} - \Gamma^0_{02}\Gamma^2_{21}$$
$$= \mathcal{O}\left(\frac{1}{r^4}\right) + \mathcal{O}\left(\frac{1}{r^5}\right) + \mathcal{O}\left(\frac{1}{r^5}\right) + \mathcal{O}\left(\frac{1}{r^4}\right) + \mathcal{O}\left(\frac{1}{r^4}\right) = \mathcal{O}\left(\frac{1}{r^4}\right)$$

$$R^0_{012} = -R^0_{021} = \partial_1\Gamma^0_{20} - \partial_2\Gamma^0_{10} + \Gamma^0_{13}\Gamma^3_{20} - \Gamma^0_{23}\Gamma^3_{01}$$
$$= \mathcal{O}\left(\frac{1}{r^4}\right) + \mathcal{O}\left(\frac{1}{r^4}\right) + \mathcal{O}\left(\frac{1}{r^5}\right) + \mathcal{O}\left(\frac{1}{r^7}\right) = \mathcal{O}\left(\frac{1}{r^4}\right)$$

$$R^1_{010} = -R^1_{001} = \partial_1\Gamma^1_{00} + \Gamma^1_{11}\Gamma^1_{00} + \Gamma^1_{12}\Gamma^2_{00} - \Gamma^1_{00}\Gamma^0_{01} - \Gamma^1_{03}\Gamma^3_{01}$$
$$= \mathcal{O}\left(\frac{1}{r^3}\right) + \mathcal{O}\left(\frac{1}{r^4}\right) + \mathcal{O}\left(\frac{1}{r^7}\right) + \mathcal{O}\left(\frac{1}{r^4}\right) + \mathcal{O}\left(\frac{1}{r^6}\right) = \mathcal{O}\left(\frac{1}{r^3}\right)$$

$$R^2_{020} = -R^0_{002} = \partial_2\Gamma^2_{00} + \Gamma^2_{21}\Gamma^1_{00} + \Gamma^2_{22}\Gamma^2_{00} - \Gamma^2_{00}\Gamma^0_{02} - \Gamma^2_{03}\Gamma^3_{02}$$
$$= \mathcal{O}\left(\frac{1}{r^5}\right) + \mathcal{O}\left(\frac{1}{r^3}\right) + \mathcal{O}\left(\frac{1}{r^7}\right) + \mathcal{O}\left(\frac{1}{r^8}\right) + \mathcal{O}\left(\frac{1}{r^6}\right) = \mathcal{O}\left(\frac{1}{r^3}\right)$$

$$R^3_{030} = -R^3_{003} = \Gamma^3_{31}\Gamma^1_{00} + \Gamma^3_{32}\Gamma^2_{00} - \Gamma^3_{01}\Gamma^1_{03} - \Gamma^3_{02}\Gamma^2_{03}$$
$$= \mathcal{O}\left(\frac{1}{r^3}\right) + \mathcal{O}\left(\frac{1}{r^5}\right) + \mathcal{O}\left(\frac{1}{r^6}\right) + \mathcal{O}\left(\frac{1}{r^6}\right) = \mathcal{O}\left(\frac{1}{r^3}\right)$$

$$R^1_{020} = -R^1_{002} = \partial_2\Gamma^1_{00} + \Gamma^1_{21}\Gamma^1_{00} + \Gamma^1_{22}\Gamma^2_{00} - \Gamma^1_{00}\Gamma^0_{02} - \Gamma^1_{03}\Gamma^3_{02}$$
$$= \mathcal{O}\left(\frac{1}{r^4}\right) + \mathcal{O}\left(\frac{1}{r^4}\right) + \mathcal{O}\left(\frac{1}{r^4}\right) + \mathcal{O}\left(\frac{1}{r^5}\right) + \mathcal{O}\left(\frac{1}{r^5}\right) = \mathcal{O}\left(\frac{1}{r^4}\right)$$

$$R^2_{010} = -R^2_{001} = \partial_1\Gamma^2_{00} + \Gamma^2_{11}\Gamma^1_{00} + \Gamma^2_{12}\Gamma^2_{00} - \Gamma^2_{00}\Gamma^0_{01} - \Gamma^2_{03}\Gamma^3_{01}$$
$$= \mathcal{O}\left(\frac{1}{r^6}\right) + \mathcal{O}\left(\frac{1}{r^6}\right) + \mathcal{O}\left(\frac{1}{r^6}\right) + \mathcal{O}\left(\frac{1}{r^7}\right) + \mathcal{O}\left(\frac{1}{r^7}\right) = \mathcal{O}\left(\frac{1}{r^6}\right)$$

$$R^0_{113} = -R^0_{131} = \partial_1\Gamma^0_{31} + \Gamma^0_{10}\Gamma^0_{31} + \Gamma^0_{13}\Gamma^3_{31} - \Gamma^0_{31}\Gamma^1_{11} - \Gamma^0_{32}\Gamma^2_{11}$$
$$= \mathcal{O}\left(\frac{1}{r^3}\right) + \mathcal{O}\left(\frac{1}{r^4}\right) + \mathcal{O}\left(\frac{1}{r^3}\right) + \mathcal{O}\left(\frac{1}{r^4}\right) + \mathcal{O}\left(\frac{1}{r^7}\right) = \mathcal{O}\left(\frac{1}{r^3}\right)$$

$$R^3_{110} = -R^3_{101} = \partial_1\Gamma^3_{01} + \Gamma^3_{10}\Gamma^0_{01} + \Gamma^3_{13}\Gamma^3_{01} - \Gamma^3_{01}\Gamma^1_{11} - \Gamma^3_{02}\Gamma^2_{11}$$
$$= \mathcal{O}\left(\frac{1}{r^5}\right) + \mathcal{O}\left(\frac{1}{r^6}\right) + \mathcal{O}\left(\frac{1}{r^5}\right) + \mathcal{O}\left(\frac{1}{r^6}\right) + \mathcal{O}\left(\frac{1}{r^7}\right) = \mathcal{O}\left(\frac{1}{r^5}\right)$$

$$R^1_{310} = -R^1_{301} = \partial_1\Gamma^1_{03} + \Gamma^1_{11}\Gamma^1_{03} + \Gamma^1_{12}\Gamma^2_{03} - \Gamma^1_{00}\Gamma^0_{31} - \Gamma^1_{03}\Gamma^3_{31}$$
$$= \mathcal{O}\left(\frac{1}{r^3}\right) + \mathcal{O}\left(\frac{1}{r^4}\right) + \mathcal{O}\left(\frac{1}{r^5}\right) + \mathcal{O}\left(\frac{1}{r^4}\right) + \mathcal{O}\left(\frac{1}{r^3}\right) = \mathcal{O}\left(\frac{1}{r^3}\right)$$

$$R^1_{130} = -R^1_{103} = \Gamma^1_{30}\Gamma^0_{01} + \Gamma^1_{33}\Gamma^3_{01} - \Gamma^1_{00}\Gamma^0_{13} - \Gamma^1_{03}\Gamma^3_{13}$$
$$= \mathcal{O}\left(\frac{1}{r^4}\right) + \mathcal{O}\left(\frac{1}{r^3}\right) + \mathcal{O}\left(\frac{1}{r^4}\right) + \mathcal{O}\left(\frac{1}{r^3}\right) = \mathcal{O}\left(\frac{1}{r^3}\right)$$

A.2. The Kerr spacetime

$$R^1_{013} = -R^1_{031} = \partial_1 \Gamma^1_{30} + \Gamma^1_{11}\Gamma^1_{30} + \Gamma^1_{12}\Gamma^2_{30} - \Gamma^1_{30}\Gamma^0_{01} - \Gamma^1_{33}\Gamma^3_{01}$$
$$= \mathcal{O}\left(\frac{1}{r^3}\right) + \mathcal{O}\left(\frac{1}{r^4}\right) + \mathcal{O}\left(\frac{1}{r^5}\right) + \mathcal{O}\left(\frac{1}{r^4}\right) + \mathcal{O}\left(\frac{1}{r^3}\right) = \mathcal{O}\left(\frac{1}{r^3}\right)$$

$$R^0_{223} = -R^0_{232} = \partial_2 \Gamma^0_{32} + \Gamma^0_{20}\Gamma^0_{32} + \Gamma^0_{23}\Gamma^3_{32} - \Gamma^0_{31}\Gamma^1_{22} - \Gamma^0_{32}\Gamma^2_{22}$$
$$= \mathcal{O}\left(\frac{1}{r^3}\right) + \mathcal{O}\left(\frac{1}{r^6}\right) + \mathcal{O}\left(\frac{1}{r^3}\right) + \mathcal{O}\left(\frac{1}{r}\right) + \mathcal{O}\left(\frac{1}{r^5}\right) = \mathcal{O}\left(\frac{1}{r}\right)$$

$$R^3_{220} = -R^3_{202} = \partial_2 \Gamma^3_{02} + \Gamma^3_{20}\Gamma^0_{02} + \Gamma^3_{23}\Gamma^3_{02} - \Gamma^3_{01}\Gamma^1_{22} - \Gamma^3_{02}\Gamma^2_{22}$$
$$= \mathcal{O}\left(\frac{1}{r^3}\right) + \mathcal{O}\left(\frac{1}{r^6}\right) + \mathcal{O}\left(\frac{1}{r^3}\right) + \mathcal{O}\left(\frac{1}{r^3}\right) + \mathcal{O}\left(\frac{1}{r^5}\right) = \mathcal{O}\left(\frac{1}{r^3}\right)$$

$$R^2_{320} = -R^2_{302} = \partial_2 \Gamma^2_{03} + \Gamma^2_{21}\Gamma^1_{03} + \Gamma^2_{22}\Gamma^2_{03} - \Gamma^2_{00}\Gamma^0_{32} - \Gamma^2_{03}\Gamma^3_{32}$$
$$= \mathcal{O}\left(\frac{1}{r^3}\right) + \mathcal{O}\left(\frac{1}{r^3}\right) + \mathcal{O}\left(\frac{1}{r^5}\right) + \mathcal{O}\left(\frac{1}{r^8}\right) + \mathcal{O}\left(\frac{1}{r^3}\right) = \mathcal{O}\left(\frac{1}{r^3}\right)$$

$$R^2_{230} = -R^2_{203} = \Gamma^2_{30}\Gamma^0_{02} + \Gamma^2_{33}\Gamma^3_{02} - \Gamma^2_{00}\Gamma^0_{23} - \Gamma^2_{03}\Gamma^3_{23}$$
$$= \mathcal{O}\left(\frac{1}{r^6}\right) + \mathcal{O}\left(\frac{1}{r^3}\right) + \mathcal{O}\left(\frac{1}{r^8}\right) + \mathcal{O}\left(\frac{1}{r^3}\right) = \mathcal{O}\left(\frac{1}{r^3}\right)$$

$$R^2_{023} = -R^2_{032} = \partial_2 \Gamma^2_{30} + \Gamma^2_{21}\Gamma^1_{30} + \Gamma^2_{22}\Gamma^2_{30} - \Gamma^2_{30}\Gamma^0_{02} - \Gamma^2_{33}\Gamma^3_{02}$$
$$= \mathcal{O}\left(\frac{1}{r^3}\right) + \mathcal{O}\left(\frac{1}{r^3}\right) + \mathcal{O}\left(\frac{1}{r^5}\right) + \mathcal{O}\left(\frac{1}{r^6}\right) + \mathcal{O}\left(\frac{1}{r^3}\right) = \mathcal{O}\left(\frac{1}{r^3}\right)$$

$$R^0_{123} = -R^0_{132} = \partial_2 \Gamma^0_{31} + \Gamma^0_{20}\Gamma^0_{31} + \Gamma^0_{23}\Gamma^3_{31} - \Gamma^0_{31}\Gamma^1_{12} - \Gamma^0_{32}\Gamma^2_{12}$$
$$= \mathcal{O}\left(\frac{1}{r^2}\right) + \mathcal{O}\left(\frac{1}{r^5}\right) + \mathcal{O}\left(\frac{1}{r^4}\right) + \mathcal{O}\left(\frac{1}{r^4}\right) + \mathcal{O}\left(\frac{1}{r^4}\right) = \mathcal{O}\left(\frac{1}{r^2}\right)$$

$$R^0_{213} = -R^0_{231} = \partial_1 \Gamma^0_{32} + \Gamma^0_{10}\Gamma^0_{32} + \Gamma^0_{13}\Gamma^3_{32} - \Gamma^0_{31}\Gamma^1_{22} - \Gamma^0_{32}\Gamma^2_{21}$$
$$= \mathcal{O}\left(\frac{1}{r^4}\right) + \mathcal{O}\left(\frac{1}{r^5}\right) + \mathcal{O}\left(\frac{1}{r^2}\right) + \mathcal{O}\left(\frac{1}{r}\right) + \mathcal{O}\left(\frac{1}{r^4}\right) = \mathcal{O}\left(\frac{1}{r}\right)$$

$$R^0_{312} = -R^0_{321} = \partial_1 \Gamma^0_{23} - \partial_2 \Gamma^0_{13} + \Gamma^0_{10}\Gamma^0_{23} + \Gamma^0_{13}\Gamma^3_{23} - \Gamma^0_{20}\Gamma^0_{31} - \Gamma^0_{23}\Gamma^3_{31}$$
$$= \mathcal{O}\left(\frac{1}{r^4}\right) + \mathcal{O}\left(\frac{1}{r^2}\right) + \mathcal{O}\left(\frac{1}{r^5}\right) + \mathcal{O}\left(\frac{1}{r^2}\right) + \mathcal{O}\left(\frac{1}{r^5}\right) + \mathcal{O}\left(\frac{1}{r^4}\right) = \mathcal{O}\left(\frac{1}{r^2}\right)$$

$$R^3_{012} = -R^3_{021} = \partial_1 \Gamma^3_{20} - \partial_2 \Gamma^3_{10} + \Gamma^3_{10}\Gamma^0_{20} + \Gamma^3_{13}\Gamma^3_{20} - \Gamma^3_{20}\Gamma^0_{01} - \Gamma^3_{23}\Gamma^3_{01}$$
$$= \mathcal{O}\left(\frac{1}{r^4}\right) + \mathcal{O}\left(\frac{1}{r^6}\right) + \mathcal{O}\left(\frac{1}{r^7}\right) + \mathcal{O}\left(\frac{1}{r^4}\right) + \mathcal{O}\left(\frac{1}{r^5}\right) + \mathcal{O}\left(\frac{1}{r^4}\right) = \mathcal{O}\left(\frac{1}{r^4}\right)$$

$$R^3_{102} = -R^3_{120} = -\partial_2 \Gamma^3_{01} + \Gamma^3_{01}\Gamma^1_{21} + \Gamma^3_{02}\Gamma^2_{21} - \Gamma^3_{20}\Gamma^0_{10} - \Gamma^3_{23}\Gamma^3_{10}$$
$$= \mathcal{O}\left(\frac{1}{r^6}\right) + \mathcal{O}\left(\frac{1}{r^6}\right) + \mathcal{O}\left(\frac{1}{r^4}\right) + \mathcal{O}\left(\frac{1}{r^5}\right) + \mathcal{O}\left(\frac{1}{r^4}\right) = \mathcal{O}\left(\frac{1}{r^4}\right)$$

$$R^3_{201} = -R^3_{210} = -\partial_1 \Gamma^3_{02} + \Gamma^3_{01}\Gamma^1_{12} + \Gamma^3_{02}\Gamma^2_{12} - \Gamma^3_{10}\Gamma^0_{20} - \Gamma^3_{13}\Gamma^3_{20}$$
$$= \mathcal{O}\left(\frac{1}{r^4}\right) + \mathcal{O}\left(\frac{1}{r^6}\right) + \mathcal{O}\left(\frac{1}{r^4}\right) + \mathcal{O}\left(\frac{1}{r^7}\right) + \mathcal{O}\left(\frac{1}{r^4}\right) = \mathcal{O}\left(\frac{1}{r^4}\right)$$

$$R^1_{023} = -R^1_{032} = \partial_2 \Gamma^1_{30} + \Gamma^1_{21}\Gamma^1_{30} + \Gamma^1_{22}\Gamma^2_{30} - \Gamma^1_{30}\Gamma^0_{02} - \Gamma^1_{33}\Gamma^3_{02}$$
$$= \mathcal{O}\left(\frac{1}{r^2}\right) + \mathcal{O}\left(\frac{1}{r^4}\right) + \mathcal{O}\left(\frac{1}{r^2}\right) + \mathcal{O}\left(\frac{1}{r^5}\right) + \mathcal{O}\left(\frac{1}{r^2}\right) = \mathcal{O}\left(\frac{1}{r^2}\right)$$

A.2. The Kerr spacetime

$$R^1_{203} = -R^1_{230} = \Gamma^1_{00}\Gamma^0_{32} + \Gamma^1_{03}\Gamma^3_{32} - \Gamma^1_{30}\Gamma^0_{20} - \Gamma^1_{33}\Gamma^3_{20}$$
$$= \mathcal{O}\left(\frac{1}{r^5}\right) + \mathcal{O}\left(\frac{1}{r^2}\right) + \mathcal{O}\left(\frac{1}{r^5}\right) + \mathcal{O}\left(\frac{1}{r^2}\right) = \mathcal{O}\left(\frac{1}{r^2}\right)$$

$$R^1_{302} = -R^1_{320} = -\partial_2\Gamma^1_{03} + \Gamma^1_{00}\Gamma^0_{23} + \Gamma^1_{03}\Gamma^3_{23} - \Gamma^1_{21}\Gamma^1_{30} - \Gamma^1_{22}\Gamma^2_{30}$$
$$= \mathcal{O}\left(\frac{1}{r^2}\right) + \mathcal{O}\left(\frac{1}{r^5}\right) + \mathcal{O}\left(\frac{1}{r^2}\right) + \mathcal{O}\left(\frac{1}{r^4}\right) + \mathcal{O}\left(\frac{1}{r^2}\right) = \mathcal{O}\left(\frac{1}{r^2}\right)$$

$$R^2_{013} = -R^2_{031} = \partial_1\Gamma^2_{30} + \Gamma^2_{11}\Gamma^1_{30} + \Gamma^2_{12}\Gamma^2_{30} - \Gamma^2_{30}\Gamma^0_{01} - \Gamma^2_{33}\Gamma^3_{01}$$
$$= \mathcal{O}\left(\frac{1}{r^4}\right) + \mathcal{O}\left(\frac{1}{r^6}\right) + \mathcal{O}\left(\frac{1}{r^4}\right) + \mathcal{O}\left(\frac{1}{r^5}\right) + \mathcal{O}\left(\frac{1}{r^4}\right) = \mathcal{O}\left(\frac{1}{r^4}\right)$$

$$R^2_{103} = -R^2_{130} = \Gamma^2_{00}\Gamma^0_{31} + \Gamma^2_{03}\Gamma^3_{31} - \Gamma^2_{30}\Gamma^0_{10} - \Gamma^2_{33}\Gamma^3_{10}$$
$$= \mathcal{O}\left(\frac{1}{r^7}\right) + \mathcal{O}\left(\frac{1}{r^4}\right) + \mathcal{O}\left(\frac{1}{r^5}\right) + \mathcal{O}\left(\frac{1}{r^4}\right) = \mathcal{O}\left(\frac{1}{r^4}\right)$$

$$R^2_{310} = -R^2_{301} = \partial_1\Gamma^2_{03} + \Gamma^2_{11}\Gamma^1_{03} + \Gamma^2_{12}\Gamma^2_{03} - \Gamma^2_{00}\Gamma^0_{31} - \Gamma^2_{03}\Gamma^3_{31}$$
$$= \mathcal{O}\left(\frac{1}{r^4}\right) + \mathcal{O}\left(\frac{1}{r^6}\right) + \mathcal{O}\left(\frac{1}{r^4}\right) + \mathcal{O}\left(\frac{1}{r^7}\right) + \mathcal{O}\left(\frac{1}{r^4}\right) = \mathcal{O}\left(\frac{1}{r^4}\right)$$

$$R^3_{330} = -R^3_{303} = \Gamma^3_{31}\Gamma^1_{03} + \Gamma^3_{32}\Gamma^2_{03} - \Gamma^3_{01}\Gamma^1_{33} - \Gamma^3_{02}\Gamma^2_{33}$$
$$= \mathcal{O}\left(\frac{1}{r^3}\right) + \mathcal{O}\left(\frac{1}{r^3}\right) + \mathcal{O}\left(\frac{1}{r^3}\right) + \mathcal{O}\left(\frac{1}{r^3}\right) = \mathcal{O}\left(\frac{1}{r^3}\right)$$

$$R^1_{112} = -R^1_{121} = \partial_1\Gamma^1_{21} - \partial_2\Gamma^1_{11} + \Gamma^1_{12}\Gamma^2_{21} - \Gamma^1_{22}\Gamma^2_{11}$$
$$= \mathcal{O}\left(\frac{1}{r^3}\right) + \mathcal{O}\left(\frac{1}{r^3}\right) + \mathcal{O}\left(\frac{1}{r^3}\right) + \mathcal{O}\left(\frac{1}{r^3}\right) = \mathcal{O}\left(\frac{1}{r^3}\right)$$

$$R^2_{221} = -R^2_{212} = \partial_2\Gamma^2_{12} - \partial_1\Gamma^2_{22} + \Gamma^2_{21}\Gamma^1_{12} - \Gamma^2_{11}\Gamma^1_{22}$$
$$= \mathcal{O}\left(\frac{1}{r^3}\right) + \mathcal{O}\left(\frac{1}{r^3}\right) + \mathcal{O}\left(\frac{1}{r^3}\right) + \mathcal{O}\left(\frac{1}{r^3}\right) = \mathcal{O}\left(\frac{1}{r^3}\right)$$

$$R^1_{212} = -R^1_{221} = \partial_1\Gamma^1_{22} - \partial_2\Gamma^1_{12} + \Gamma^1_{11}\Gamma^1_{22} + \Gamma^1_{12}\Gamma^2_{22} - \Gamma^1_{21}\Gamma^1_{21} - \Gamma^1_{22}\Gamma^2_{21}$$
$$= \text{const.} + \mathcal{O}\left(\frac{1}{r^2}\right) + \mathcal{O}\left(\frac{1}{r}\right) + \mathcal{O}\left(\frac{1}{r^4}\right) + \mathcal{O}\left(\frac{1}{r^4}\right) + \text{const.} = \text{const.}$$

$$R^2_{121} = -R^2_{112} = \partial_2\Gamma^2_{11} - \partial_1\Gamma^2_{21} + \Gamma^2_{21}\Gamma^1_{11} + \Gamma^2_{22}\Gamma^2_{11} - \Gamma^2_{11}\Gamma^1_{12} - \Gamma^2_{12}\Gamma^2_{12}$$
$$= \mathcal{O}\left(\frac{1}{r^4}\right) + \mathcal{O}\left(\frac{1}{r^2}\right) + \mathcal{O}\left(\frac{1}{r^3}\right) + \mathcal{O}\left(\frac{1}{r^6}\right) + \mathcal{O}\left(\frac{1}{r^6}\right) + \mathcal{O}\left(\frac{1}{r^2}\right) = \mathcal{O}\left(\frac{1}{r^2}\right)$$

$$R^1_{313} = -R^1_{331} = \partial_1\Gamma^1_{33} + \Gamma^1_{11}\Gamma^1_{33} + \Gamma^1_{12}\Gamma^2_{33} - \Gamma^1_{30}\Gamma^0_{31} - \Gamma^1_{33}\Gamma^3_{31}$$
$$= \text{const.} + \mathcal{O}\left(\frac{1}{r}\right) + \mathcal{O}\left(\frac{1}{r^2}\right) + \mathcal{O}\left(\frac{1}{r^4}\right) + \text{const.} = \text{const.}$$

$$R^3_{131} = -R^3_{113} = -\partial_1\Gamma^3_{31} + \Gamma^3_{31}\Gamma^1_{11} + \Gamma^3_{32}\Gamma^2_{11} - \Gamma^3_{10}\Gamma^0_{13} - \Gamma^3_{13}\Gamma^3_{13}$$
$$= \mathcal{O}\left(\frac{1}{r^2}\right) + \mathcal{O}\left(\frac{1}{r^3}\right) + \mathcal{O}\left(\frac{1}{r^4}\right) + \mathcal{O}\left(\frac{1}{r^6}\right) + \mathcal{O}\left(\frac{1}{r^2}\right) = \mathcal{O}\left(\frac{1}{r^2}\right)$$

$$R^2_{323} = -R^2_{332} = \partial_2\Gamma^2_{33} + \Gamma^2_{21}\Gamma^1_{33} + \Gamma^2_{22}\Gamma^2_{33} - \Gamma^2_{30}\Gamma^0_{32} - \Gamma^2_{33}\Gamma^3_{32}$$
$$= \text{const.} + \text{const.} + \mathcal{O}\left(\frac{1}{r^2}\right) + \mathcal{O}\left(\frac{1}{r^6}\right) + \text{const.} = \text{const.}$$

$$R^3_{232} = -R^3_{223} = -\partial_2\Gamma^3_{32} + \Gamma^3_{31}\Gamma^1_{22} + \Gamma^3_{32}\Gamma^2_{22} - \Gamma^3_{20}\Gamma^0_{23} - \Gamma^3_{23}\Gamma^3_{23}$$
$$= \text{const.} + \text{const.} + \mathcal{O}\left(\frac{1}{r^2}\right) + \mathcal{O}\left(\frac{1}{r^6}\right) + \text{const.} = \text{const.}$$

$$R^3_{312} = -R^3_{321} = \partial_1\Gamma^3_{23} - \partial_2\Gamma^3_{13} + \Gamma^3_{10}\Gamma^0_{23} - \Gamma^3_{20}\Gamma^0_{31}$$
$$= \mathcal{O}\left(\frac{1}{r^4}\right) + \mathcal{O}\left(\frac{1}{r^4}\right) + \mathcal{O}\left(\frac{1}{r^7}\right) + \mathcal{O}\left(\frac{1}{r^5}\right) = \mathcal{O}\left(\frac{1}{r^4}\right)$$

$$R^3_{132} = -R^3_{123} = -\partial_2\Gamma^3_{31} + \Gamma^3_{31}\Gamma^1_{21} + \Gamma^3_{32}\Gamma^2_{21} - \Gamma^3_{20}\Gamma^0_{13} - \Gamma^3_{23}\Gamma^3_{13}$$
$$= \mathcal{O}\left(\frac{1}{r^4}\right) + \mathcal{O}\left(\frac{1}{r^3}\right) + \mathcal{O}\left(\frac{1}{r}\right) + \mathcal{O}\left(\frac{1}{r^5}\right) + \mathcal{O}\left(\frac{1}{r}\right) = \mathcal{O}\left(\frac{1}{r}\right)$$

$$R^3_{231} = -R^3_{213} = -\partial_1\Gamma^3_{32} + \Gamma^3_{31}\Gamma^1_{12} + \Gamma^3_{32}\Gamma^2_{12} - \Gamma^3_{10}\Gamma^0_{23} - \Gamma^3_{13}\Gamma^3_{23}$$
$$= \mathcal{O}\left(\frac{1}{r^4}\right) + \mathcal{O}\left(\frac{1}{r^3}\right) + \mathcal{O}\left(\frac{1}{r}\right) + \mathcal{O}\left(\frac{1}{r^7}\right) + \mathcal{O}\left(\frac{1}{r}\right) = \mathcal{O}\left(\frac{1}{r}\right)$$

$$R^1_{332} = -R^1_{323} = -\partial_2\Gamma^1_{33} + \Gamma^1_{30}\Gamma^0_{23} + \Gamma^1_{33}\Gamma^3_{23} - \Gamma^1_{21}\Gamma^1_{33} - \Gamma^1_{22}\Gamma^2_{33}$$
$$= \mathcal{O}(r) + \mathcal{O}\left(\frac{1}{r^5}\right) + \mathcal{O}(r) + \mathcal{O}\left(\frac{1}{r}\right) + \mathcal{O}(r) = \mathcal{O}(r)$$

$$R^2_{331} = -R^2_{313} = -\partial_1\Gamma^2_{33} + \Gamma^2_{30}\Gamma^0_{13} + \Gamma^2_{33}\Gamma^3_{13} - \Gamma^2_{11}\Gamma^1_{33} - \Gamma^2_{12}\Gamma^2_{33}$$
$$= \mathcal{O}\left(\frac{1}{r^2}\right) + \mathcal{O}\left(\frac{1}{r^5}\right) + \mathcal{O}\left(\frac{1}{r}\right) + \mathcal{O}\left(\frac{1}{r^3}\right) + \mathcal{O}\left(\frac{1}{r}\right) = \mathcal{O}\left(\frac{1}{r}\right)$$

As it is our aim to show that the vanishing behaviour of the components of the curvature tensor is of order $\mathcal{O}\left(\frac{1}{r^3}\right)$, we have to calculate some of the components more precisely. We then get:

$$R^1_{212} : \partial_1\Gamma^1_{22} - \Gamma^1_{22}\Gamma^2_{21} = -\frac{r^4}{\rho^2} + \mathcal{O}\left(\frac{1}{r}\right) + \frac{\tilde{\Delta}r}{\rho} \cdot \frac{r}{\rho} = -\frac{r^4}{\rho^2} + \frac{r^4}{\rho^2} + \mathcal{O}\left(\frac{1}{r}\right) = \mathcal{O}\left(\frac{1}{r}\right)$$
$$\Rightarrow R^1_{212} = \mathcal{O}\left(\frac{1}{r}\right)$$

$$R^2_{121} : -\partial_1\Gamma^2_{21} - \Gamma^2_{12}\Gamma^2_{12} = \frac{1}{\rho^2}(r^2 - a^2\cos^2\vartheta) - \frac{r^2}{\rho^2} = -\frac{a^2\cos^2\vartheta}{\rho^2} = \mathcal{O}\left(\frac{1}{r^4}\right)$$
$$\Rightarrow R^2_{121} = \mathcal{O}\left(\frac{1}{r^3}\right)$$

$$R^1_{313} : \partial_1\Gamma^1_{33} - \Gamma^1_{33}\Gamma^3_{31} = -\sin^2\vartheta\frac{r^4}{\rho^2} + \mathcal{O}\left(\frac{1}{r}\right) + \left(\sin^2\vartheta\frac{\tilde{\Delta}r}{\rho} + \mathcal{O}\left(\frac{1}{r^2}\right)\right)\left(\frac{r}{\tilde{\Delta}} + \mathcal{O}\left(\frac{1}{r^2}\right)\right)$$
$$= -\sin^2\vartheta \cdot \frac{r^4}{\rho^2} + \sin^2\vartheta \cdot \frac{r^4}{\rho^2} + \mathcal{O}\left(\frac{1}{r}\right) = \mathcal{O}\left(\frac{1}{r}\right) \Rightarrow R^1_{313} = \mathcal{O}\left(\frac{1}{r}\right)$$

$$R^3_{131} : -\partial_1\Gamma^3_{31} - \Gamma^3_{13}\Gamma^3_{13} = -\frac{-r^2 + a^2}{\tilde{\Delta}^2} + \mathcal{O}\left(\frac{1}{r^3}\right) - \frac{r^2}{\tilde{\Delta}^2} + \mathcal{O}\left(\frac{1}{r^3}\right) = \mathcal{O}\left(\frac{1}{r^3}\right)$$
$$\Rightarrow R^3_{131} = \mathcal{O}\left(\frac{1}{r^3}\right)$$

A.2. The Kerr spacetime

$$R^2_{323} : \partial_2 \Gamma^2_{33} + \Gamma^1_{21}\Gamma^1_{33} - \Gamma^2_{33}\Gamma^3_{32} = -\frac{r^2(\cos^2 \vartheta - \sin^2 \vartheta)}{\rho} + \mathcal{O}\left(\frac{1}{r}\right) + \frac{r}{\rho}(-\sin^2 \vartheta + \text{const.})$$

$$- \left(-\sin \vartheta \cos \vartheta \cdot \frac{r^2}{\rho} + \mathcal{O}\left(\frac{1}{r}\right)\right) \cdot \left(\frac{\cos \vartheta}{\sin \vartheta} + \mathcal{O}\left(\frac{1}{r^3}\right)\right)$$

$$= \frac{r^2}{\rho}\left(-\cos^2 \vartheta + \sin^2 \vartheta - \sin^2 \vartheta + \sin \vartheta \cos \vartheta \cdot \frac{\cos \vartheta}{\sin \vartheta}\right) + \mathcal{O}\left(\frac{1}{r}\right) = \mathcal{O}\left(\frac{1}{r}\right)$$

$$\Rightarrow R^2_{323} = \mathcal{O}\left(\frac{1}{r}\right)$$

$$R^3_{232} : -\partial_2 \Gamma^3_{32} + \Gamma^3_{31}\Gamma^1_{22} - \Gamma^3_{23}\Gamma^3_{23}$$

$$= \frac{1}{\sin^2 \vartheta} + \mathcal{O}\left(\frac{1}{r^3}\right) + \left(\frac{r}{\widetilde{\Delta}} + \mathcal{O}\left(\frac{1}{r^2}\right)\right) \cdot \left(-\frac{\widetilde{\Delta} r}{\rho}\right) + \frac{\cos^2 \vartheta}{\sin^2 \vartheta} + \mathcal{O}\left(\frac{1}{r^3}\right)$$

$$= 1 - 1 + \frac{a^2 \cos^2 \vartheta}{\rho^2} + \mathcal{O}\left(\frac{1}{r}\right) = \mathcal{O}\left(\frac{1}{r}\right) \Rightarrow R^3_{232} = \mathcal{O}\left(\frac{1}{r}\right)$$

$$R^3_{132}/R^3_{231} : \Gamma^3_{32}\Gamma^2_{21} - \Gamma^3_{23}\Gamma^3_{13} = \left(\frac{\cos \vartheta}{\sin \vartheta} + \mathcal{O}\left(\frac{1}{r^3}\right)\right) \cdot \left(\frac{r}{\rho} - \frac{r}{\widetilde{\Delta}} + \mathcal{O}\left(\frac{1}{r^2}\right)\right)$$

$$= \left(\frac{\cos \vartheta}{\sin \vartheta} + \mathcal{O}\left(\frac{1}{r^3}\right)\right) \cdot \left(\frac{r^3 - r^3}{\widetilde{\Delta}\rho} + \mathcal{O}\left(\frac{1}{r^2}\right)\right) = \mathcal{O}\left(\frac{1}{r^2}\right) \Rightarrow R^3_{132}/R^3_{231} = \mathcal{O}\left(\frac{1}{r^2}\right)$$

$$R^1_{332} : -\partial_2 \Gamma^1_{33} + \Gamma^1_{33}\Gamma^3_{23} - \Gamma^1_{22}\Gamma^2_{33}$$

$$= \frac{2\widetilde{\Delta} r \sin \vartheta \cos \vartheta}{\rho} + \mathcal{O}\left(\frac{1}{r}\right) + \left(-\frac{\widetilde{\Delta} r \sin^2 \vartheta}{\rho} + \mathcal{O}\left(\frac{1}{r^2}\right)\right) \cdot \left(\left(\frac{\cos \vartheta}{\sin \vartheta}\right) + \mathcal{O}\left(\frac{1}{r^3}\right)\right)$$

$$- \left(-\frac{\widetilde{\Delta} r}{\rho}\right) \cdot \left(-\frac{\widetilde{\Delta} \sin \vartheta \cos \vartheta}{\rho} + \mathcal{O}\left(\frac{1}{r}\right)\right)$$

$$= \frac{2r^5 \sin \vartheta \cos \vartheta}{\rho^2} - \frac{r^5 \sin \vartheta \cos \vartheta}{\rho^2} - \frac{r^5 \sin \vartheta \cos \vartheta}{\rho^2} + \text{const.} = \text{const.}$$

$$\Rightarrow R^1_{332} = \text{const.}$$

$$R^2_{331} : \Gamma^3_{33}\Gamma^3_{13} - \Gamma^2_{12}\Gamma^2_{33} = \left(\frac{\cos \vartheta}{\sin \vartheta} + \mathcal{O}\left(\frac{1}{r^3}\right)\right) \cdot \left(\frac{r}{\widetilde{\Delta}} - \frac{r}{\rho}\right)$$

$$= \left(\frac{\cos \vartheta}{\sin \vartheta} + \mathcal{O}\left(\frac{1}{r^3}\right)\right) \cdot \left(\frac{r^3 - r^3}{\widetilde{\Delta}\rho} + \mathcal{O}\left(\frac{1}{r^2}\right)\right) = \mathcal{O}\left(\frac{1}{r^2}\right) \Rightarrow R^2_{331} = \mathcal{O}\left(\frac{1}{r^2}\right).$$

Index

Absolute time, 48
Annihilator of a vector, 14
Annihilator of tangent vector, 21
Axioms of frame theory, 21

Ball around zero, 130
Bilinear form
    negative definite, 14, 15
    negative semi-definite, 14, 15
    positive definite, 14, 15
    positive semi-definite, 14, 15
    indefinite, 14, 15
    non-degenerate, 15

Causality constant, 20
Christoffel symbol, 12
Christoffel symbols
    Transformation behaviour, 116
Classical Newtonian system, 75
    Imbedding, 76
Co-normal, 129
Coordinate system
    adapted, 115
Curvature tensor, 9
    anti-symmetry, 9
    First Bianchi-identity, 9
    Second Bianchi-identity, 11
Curve
    spacelike, 60
Cylinder around zero, 130

Degenerate cone, 92
Distribution, 40
    integrable, 40
    involutive, 40
Divergence, 13

Ehlers spacetime, 75
Einstein curvature tensor, 13
Einstein spacetime, 5
Einstein vacuum equation, 19
Einstein's field equations, 19
Einstein's law of motion, 18
Energy momentum tensor, 19

Euler equations, 18, 72
Extension of spacetime, 79

Family of Ehlers spacetimes, 75
    asymptotically flat at spacelike infinity, 132
Force, 17
Formula of Cartan, 43
free falling particle, 19

Galiläi transformation, 124
Gravitation field, 18
Gravity constant, 25
Gravity field, 21

Hypersurface
    spacelike, 130

Index of bilinear form, 15
Integral manifold, 40

Kerr metric, 85, 91
    extended, 86, 91
Kerr Spacetime, 140, 158
Kerr spacetime, 85
Killing vector field, 101

Lie derivation, 100
Lorentzian manifold, 16
    time-oriented, 17
Lorentzian metric, 16
Lorenzian metric, 130
    parametrized, 131

Manifold
    geodesically complete in space direction, 61
Mass density, 18
mass density , 18
Matter tensor, 21
Minkowski metric, 80, 90
    extended, 80, 90
Minkowski scalar product, 79
Minkowski space, 79
Minkowski spacetime, 79, 90, 134

Natural pairing between covectors and tangent vectors, 8
Newtonian law of gravitation, 18
Newtonian limit, 78
Newtonian particle, 17
Newtonian space, 17
Newtonian spacetime, 17
Newtonian time, 17
Normal form theorem, 29
Normal vector, 130

Observer, 101
    irrotational, 101

Perfect fluid, 18
Pressure function, 18

Quasi-Newtonian equations, 73
Quasi-Newtonian limit, 77
Quasi-Newtonian system, 77
    Imbedding, 77

Rank of bilinear form, 15
Ricci tensor, 11
Riemannian curvature tensor, 10
    First Bianchi-identity, 10

Scalar curvature, 11
Schwarzschild metric, 81, 90
    extended, 82, 90
Schwarzschild spacetime, 81, 90, 134, 148
Second Bianchi-identity
    Contracted, 13
Section
    global, 7
Space metric, 20
Spaceleaf, 130
Spacetime, 17
    asymptotically flat at spacelike infinity, 130
    spherically symmetric, 104
    static, 102
    stationary, 102
Submersion, 44

Tangent vector
    timelike, 21
Theorem of Frobenius, 40
Theorem of Sylvester, 15
Time metric, 20
Time-orientation, 17
Transitions between tangent and cotangent space, 7

Vacuum, 19
Vector
    null, 17
    spacelike, 17
    timelike, 16
Vector field
    equivalent, 17
    future-oriented, 17
    past-oriented, 17
Velocity field, 18

Bibliography

[1] T. W. BAUMGARTE, *The Newtonian Limit in a Model Problem*, General Relativity and Gravitation, 25 (1993), pp. 1189–1204.

[2] N. BERLINE, E. GETZLER, AND M. VERGNE, *Heat Kernels and Dirac Operators*, Berlin, 1992.

[3] G. E. BREDON, *Introduction to compact transformation groups*, New York, 1972.

[4] E. CARTAN, *Sur les Variétés a Connexion Affine et la Théorie de la Relativité Généralisée*, Annales Scientifiques de L'École Normale Supérieure, 58 (1923), pp. 325–412.

[5] E. CARTAN, *Sur les Variétés a Connexion Affine et la Théorie de la Relativité Généralisée (Suite)*, Annales Scientifiques de L'École Normale Supérieure, 59 (1924), pp. 1–25.

[6] I. CHAVEL, *Riemannian geometry: A modern introduction*, Cambridge, 1993.

[7] M. P. DO CARMO, *Riemannian Geometry*, Boston, 1992.

[8] J. EHLERS, *The Nature and Structure of Spacetime*, in The Physicist's Conception of Nature: Symposium on the Development of the Physicst's Conception of Nature in the 20. century; held at the Internat. Centre for Theoretical Physics, Miramare, Trieste, Italy, 18 - 25 sept. 1972, J. Mehra, ed., Dordrecht, 1973, pp. 71–91.

[9] J. EHLERS, *Isolated Systems in General Relativity*, in Ninth Texas Symposium on relativistic Astrophysics (Annals of the New York Academy of Science, Volume 336), J. Ehlers, J. J. Perry, and M. Walker, eds., New York, 1980, pp. 279–294.

[10] J. EHLERS, *Über den Newtonschen Grenzwert der Einsteinschen Gravitationstheorie*, in Grundlagenprobleme der modernen Physik: Festschrift für Peter Mittelstaedt zum 50. Geburtstag, Nitsch, Pfarr, and Stachow, eds., 1981, pp. 65–84.

[11] J. EHLERS, *On the limit relations between, and approximative explanations of, physical theories*, in Logic, Methodology and Philosophy of Science VII. Proceedings of the seventh international congress of Logic, Methodology and Philosophy of Science, Salzburg, 1983, R. Barcan Marcus, G. Dorn, and P. Weingartner, eds., 1986, pp. 388–403.

[12] J. EHLERS, *The Newtonian Limit of General Relativity*, in Classical Mechanics and Relativity: Relationship and Consistency, G. Ferrarese, ed., 1991, pp. 95–106.

[13] J. EHLERS, *Examples of Newtonian Limits of relativistic spacetimes*, Classical and Quantum Gravity, 14 (1997), pp. A119–A126.

[14] J. EHLERS, *The Newtonian Limit of General Relativity*, in Understanding Physics. Copernicus Gesellschaft e.V. Katlenburg, A. K. Richter, ed., Lindau, 1998, pp. 1–13.

[15] C. EHRESMANN, *Les connexions infinitésimales dans un espace fibré différentiable*, Colloque de Topologie (Espaces Fibrés), tenu à Bruxelles du 5 au 8 juin 1950 (Publications du Centre Belge de Recherches Mathématiques; 2), (1950), pp. 28–55.

[16] A. EINSTEIN, *Die Grundlage der allgemeinen Relativitätstheorie*, Annalen der Physik. Vierte Folge, 49 (1916), pp. 769–822.

[17] G. FISCHER, *Lineare Algebra*, Braunschweig/Wiesbaden, 2000.

[18] H. FISCHER AND H. KAUL, *Mathematik für Physiker*, vol. 3: Variationsrechnung, Differentialgeometrie, Mathematische Grundlagen der allgemeinen Relativitätstheorie, Wiesbaden, 2 ed., 2006.

[19] K. FRIEDRICHS, *Eine invariante Formulierung des Newtonschen Gravitationsgesetzes und des Grenzüberganges vom Einsteinschen zum Newtonschen Gesetz*, Mathematische Annalen, 98 (1928), pp. 566–575.

[20] R. GEROCH, *Asymptotic structure of space-time*, in Asymptotic structure of space-time: proceedings of a Symposium on Asymptotic Structure of Space-Time, held at Univ. of Cincinnati, Ohio, June 14 - 18, 1976, F. P. Esposito, ed., 1977, pp. 1–105.

[21] R. GREVE, *Kontinuumsmechanik. Ein Grundkurs*, Berlin, 2003.

[22] A. HATCHER, *Algebraic topology*, Cambridge, 2006.

[23] A. HATCHER, *Vector bundles and k-theory*. http://www.math.cornell.edu/ ~hatcher/, 2009.

[24] P. HAVAS, *Four-Dimensional Formulations of Newtonian Mechanics and Their Relation to the Special and the General Theory of Relativity*, Reviews of Modern Physics, 36 (1964), pp. 938–965.

[25] S. W. HAWKING AND G. F. R. ELLIS, *The large scale structure of space-time*, vol. 1, Cambridge monographs on mathematical physics, Cambridge, 1974.

[26] U. HEILIG, *On the Existence of Rotating Stars in General Relativity*, Communications in Mathematical Physics, 166 (1995), pp. 457–493.

[27] K. JAENICH, *Differnzierbare G-Mannigfaltigkeiten (Lecture notes in mathematics, 59)*, Berlin, 1968.

[28] J. JOST, *Riemannian Geometry and Geometric Analysis*, Berlin, 2008.

[29] H. KÜNZLE, *Galilei and Lorentz structures on space-time: comparison of the corresponding geometry and physics*, Annales de l'I.H.P., 17 (1972), pp. 337–362.

[30] H. KÜNZLE, *Covariant Newtonian Limit of Lorentz Space-Times*, General Relativity and Gravitation, 7 (1976), pp. 445–457.

[31] M. LOTTERMOSER, *Über den Newtonschen Grenzwert der Allgemeinen Relativitätstheorie und die relativistische Erweiterung Newtonscher Anfangsdaten*, PhD thesis, Fakultät für Physik der Ludwig-Maximilians-Universität München, 1988.

[32] T. A. OLIYNYK, *The Newtonian limit for perfect fluids*, Communications in Mathematical Physics, 276 (2007), pp. 131–188.

[33] T. A. OLIYNYK AND B. SCHMIDT, *Existence of families of spacetimes with a Newtonian limit*, General Relativity and Gravitation, 41 (2009), pp. 2093–2111.

[34] R. OLOFF, *Geometrie der Raumzeit. Eine mathematische Einführung in die Relativitätstheorie*, Wiesbaden, 3 ed., 2004.

[35] B. O'NEILL, *Semi-Riemannian geometry. With application to relativity*, San Diego, 1993.

[36] B. O'NEILL, *The Geometry of Kerr Black Holes*, Wellesley, 1996.

[37] P. PETERSEN, *Riemannian Geometry*, New York, 1998.

[38] A. D. RENDALL, *The Newtonian Limit for Asymptotically Flat Solutions of the Vlasov-Einstein System*, Communications in Mathematical Physics, 163 (1994), pp. 89–112.

[39] R. SCHULMANN, J. KOX, M. JANSSEN, AND J. ILLY, eds., *The collected papers of Albert Einstein, Volume 8: The Berlin Years: Correspondence, 1914-1918, Part A:1914-1917*, Princeton, 1998.

[40] E. H. SPANIER, *Algebraic Topology*, New York, 1966.

[41] N. STRAUMANN, *General Relativity. With Applications to Astrophyics*, Berlin, 2004.

[42] A. TRAUTMAN, *Comparison of Newtonian and Relativistic Theories of Space-Time*, in Perspectives in Geometry and Relativity. Essays in Honor of Václav Hlavatý, B. Hoffmann, ed., 1966, pp. 413–425.

[43] R. M. WALD, *General Relativity*, Chicago, 1984.

[44] F. W. WARNER, *Foundations of Differentiable Manifolds and Lie Groups*, New York, 1983.

I want morebooks!

Buy your books fast and straightforward online - at one of world's fastest growing online book stores! Environmentally sound due to Print-on-Demand technologies.

Buy your books online at
www.morebooks.shop

Kaufen Sie Ihre Bücher schnell und unkompliziert online – auf einer der am schnellsten wachsenden Buchhandelsplattformen weltweit! Dank Print-On-Demand umwelt- und ressourcenschonend produziert.

Bücher schneller online kaufen
www.morebooks.shop

KS OmniScriptum Publishing
Brivibas gatve 197
LV-1039 Riga, Latvia
Telefax +371 686 204 55

info@omniscriptum.com
www.omniscriptum.com

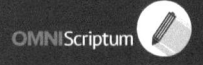

Printed by Books on Demand GmbH, Norderstedt / Germany